Autonomous Vehicle and Smart Traffic

Edited by Sezgin Ersoy and Tayyab Waqar

Published in London, United Kingdom

IntechOpen

Supporting open minds since 2005

Autonomous Vehicle and Smart Traffic
http://dx.doi.org/10.5772/intechopen.81968
Edited by Sezgin Ersoy and Tayyab Waqar

Contributors
Zia Ur Rahman Farooqi, Muhammad Sabir, Nukshab Zeeshan, Ghulam Murtaza, Muhammad Mahroz Hussian, Muhammad Usman Ghani, Sylvie Mira Bonnardel, Danielle Attias, Fabio Antonialli, Ayesha Iqbal, Fatma Ben Salem, Ishak Ertugrul, Osman Ulkir, Sezgin Ersoy, Tayyab Waqar, Mustafa F. S. Zortul

First published in London, United Kingdom, 2020 by IntechOpen
IntechOpen is the global imprint of INTECHOPEN LIMITED, registered in England and Wales, registration number: 11086078, 7th floor, 10 Lower Thames Street, London, EC3R 6AF, United Kingdom
Printed in Croatia

British Library Cataloguing-in-Publication Data
A catalogue record for this book is available from the British Library

Additional hard and PDF copies can be obtained from orders@intechopen.com

Autonomous Vehicle and Smart Traffic
Edited by Sezgin Ersoy and Tayyab Waqar
p. cm.
Print ISBN 978-1-78984-337-8
Online ISBN 978-1-78984-338-5
eBook (PDF) ISBN 978-1-83968-397-8

An electronic version of this book is freely available, thanks to the support of libraries working with Knowledge Unlatched. KU is a collaborative initiative designed to make high quality books Open Access for the public good. More information about the initiative and links to the Open Access version can be found at www.knowledgeunlatched.org

Meet the editors

Sezgin Ersoy is an associate professor of mechatronics engineering and material science. After graduating from Marmara University, he became a faculty member at the same university. His publications include a variety of efforts to understand changes in automotive mechatronics, polymer science, and biomedical technologies. He was granted fellowship at the TUBİTAK at Bourgogne University ISAT and spent one year as a visiting fellow there to study several projects during 2014/2015. He is the author of a chapter "Science Education in a Rapidly Changing World", USA, 2011, and coauthor of "*Acoustic Properties of Bio-Materials*", Stuttgart, 2010. He has two national science awards and is an editorial member of several scientific journals.

Tayyab Waqar is currently pursuing his PhD at Marmara University, Turkey, in the Mechatronics Engineering Department. His PhD work is based on the designing, characterization, and fabrication studies of surface acoustic wave-based sensors. At the same time, he is also working as an R&D specialist at Arçelik/Beko group in the Sensor Technologies/Mechatronics R&D Department. He is also performing duties as a reviewer for IEEE, Elsevier, and Wiley. He received his Master of Engineering degree in Mechatronics Engineering in 2015 from Marmara University in Turkey. His master's work was based on fault diagnosis in gears using artificial intelligence. During his master's, he worked as a research assistant at Marmara University. He completed his Bachelor of Engineering degree from Dawood College of Engineering and Technology, Pakistan, in Electronics Engineering in 2012. During his bachelor studies, he worked on control and automation engineering for industrial manufacturing.

Contents

Preface

Efforts are being made globally by policymakers and leaders to make the world we live in more efficient, in every part of life, by introducing smart systems under the concept of smart cities. Autonomous transport and cars are also an important part of those approaches.

There are six levels of automation for transport as defined by SAE International and J3016 in 2014. Level 0 means no automation whatsoever and level 5 defines fully autonomous vehicles that require no human input at all. These levels are further explained in the table below.

SAE level	Name	Narrative Definition	Execution of Steering and Acceleration/ Deceleration	*Monitoring* of Driving Environment	Fallback Performance of *Dynamic Driving Task*	System Capability *(Driving Modes)*
***Human driver* monitors the driving environment**						
0	**No Automation**	the full-time performance by the *human driver* of all aspects of the *dynamic driving task*, even when enhanced by warning or intervention systems	Human driver	Human driver	Human driver	n/a
1	**Driver Assistance**	the *driving mode*-specific execution by a driver assistance system of either steering or acceleration/deceleration using information about the driving environment and with the expectation that the *human driver* perform all remaining aspects of the *dynamic driving task*	Human driver and system	Human driver	Human driver	Some driving modes
2	**Partial Automation**	the *driving mode*-specific execution by one or more driver assistance systems of both steering and acceleration/ deceleration using information about the driving environment and with the expectation that the *human driver* perform all remaining aspects of the *dynamic driving task*	**System**	Human driver	Human driver	Some driving modes
***Automated driving system* ("system") monitors the driving environment**						
3	**Conditional Automation**	the *driving mode*-specific performance by an *automated driving system* of all aspects of the dynamic driving task with the expectation that the *human driver* will respond appropriately to a *request to intervene*	System	**System**	Human driver	Some driving modes
4	**High Automation**	the *driving mode*-specific performance by an automated driving system of all aspects of the *dynamic driving task*, even if a *human driver* does not respond appropriately to a *request to intervene*	System	System	**System**	Some driving modes
5	**Full Automation**	the full-time performance by an *automated driving system* of all aspects of the *dynamic driving task* under all roadway and environmental conditions that can be managed by a *human driver*	System	System	System	**All driving modes**

The introduction of autonomous vehicles, level 5, will not only make our everyday traveling much safer but it will also help us to curb pollution and traffic congestion while adding more value by allowing us to increase our productivity during our travels. The increase in the use of autonomous vehicles will also help in the elimination of human error, thus decreasing traffic accidents due to reckless driving and lack of attention. Self-driving vehicles will also enable mobility for those who for different reasons either are not allowed to or cannot drive a car. It will increase their life quality by allowing them faster access to their daily routines without depending on a driver. Since autonomous vehicle technology will be based on electricity, it will help in limiting the pollution caused by vehicles. It may also introduce the concept of ride-sharing, where one vehicle can be used by more than one person at the same time.

Of course, certain issues need to be taken care of to sustain the aforementioned benefits of autonomous vehicles. For example:

- It seems highly impossible at the beginning for all vehicles on the road to be at level 5 of automation. They will be faced with sharing the road with vehicles that will have lower levels of automation.

- Another issue will be the security of those vehicles. There is a risk of those vehicles being hacked and manipulated for malicious activity. This is something that needs to be managed very carefully.

- Although autonomous driving technology will most likely be based on electricity, which will make it environment friendly, it will still not help in eliminating pollution because fossil fuels are still a major resource for generating electricity.

- The use of sophisticated technologies in the vehicle will require expertise during the repair process and hence it will be more costly to repair in case of a breakdown.

There is no doubt that autonomous transport will be an essential part of the future smart city concept's implementation. To this end, policymakers and professionals have a major role to play in the successful development and implementation of autonomous vehicle concepts.

Sezgin Ersoy
Marmara University,
Turkey

Tayyab Waqar
Arcelik AŞ,
Turkey

Section 1

Environmental Factors

Chapter 1

Vehicular Noise Pollution: Its Environmental Implications and Strategic Control

Zia Ur Rahman Farooqi, Muhammad Sabir, Nukshab Zeeshan, Ghulam Murtaza, Muhammad Mahroz Hussain and Muhammad Usman Ghani

Abstract

Noise pollution has been recognized as one of the major hazard that impacts the quality of life all around the world. Because of the rapid increase in technology, industrialization, urbanization and other communication and transport systems, noise pollution has reached to a disturbing level over the years which needs to be studied and controlled to avoid different health effects like high blood pressure, sleeplessness, nausea, heart attack, depression, dizziness, headache, and induced hearing loss. To address this situation, different countries have different strategies like vehicular noise limits and their regulation, vehicles physical health checkup, different time of operations for noisy traffic like trucks in the evening or night time, and noise pollution fines for noisy vehicles.

Keywords: traffic noise, environment, community effects, human health

1. Introduction

A sound is termed as noise when it becomes continuous and above the threshold limits of the ears [1]. A vehicular noise is the resultant of the vibrating body of the vehicle plus its engine operating sound [2]. Noise has different types including impulsive noise, continuous noise, intermittent noise and low frequency noise. All above mentioned types of noise are dangerous to human and animals if their limits are exceeded [3]. Noise affects people so badly that at some places policy makers were compelled to said that there should be restrictions on noisy vehicles for reducing noise pollution as in New Delhi, India [4] and Guangzhou district, China [5]. Noise is a common problem in urban areas as compared to the villages because of the mechanization and more vehicles on the road [6, 7]. All types of noise altogether affect the same irrespective of the sources and cause headache to the high blood pressure and other heart diseases [8]. Some noise types like aircraft and train noise also has the negative effects on property prices [9, 10]. The noises from cracks in a refrigerator, fan and air conditioner affects evenly as a loud noise of train or aircraft because of its continuity. These results were obtained by a study that the noises in household appliances operations are one of the main sources of noise [11]. It is one of the big problems in industrial areas. It is produced from mechanical operations, transportation vehicles and different sirens within the industry [12]. Noise also disturbs mining worker and their operations [13]. Due to this noise, workers face different health issues including hearing loss, dizziness, headaches, high blood pressure and anxiety [14, 15]. It is a fact that noise in the vicinity of airports is a public health issue and its exposure affect sleep quality, restlessness and headache [16]. This noise also increases vascular and cerebral oxidative stress and triggers vascular dysfunction [17, 18]. In addition to the above, cargo ships are also the source of nuisance and sleep disturbance. However, these are usually not sources of noise as they affect only small communities living in harbors [19]. Low frequency noise is also included in the noise types and effects humans by annoyance and sleep disturbances [20]. Migraines, tinnitus, nausea, sleep disorders, insomnia, quality of life and minor stress strokes are the result of low frequency noise [21]. Household and community noise is also a health issue. As we live in houses for the major part of the day and we are exposed continuously to different noise types like lawn mower noise, dogs barking, kitchen grinder operation and sound systems/ television which seems dangerous. It is given in the documented form that there is an association with several diseases and the growing number of exposed persons all over the world with this type of noise. The effects are ranged from cardiovascular diseases to metabolic disorders [22, 23]. It also has impacts on animals including frogs to the whales and elephants by affecting their reproduction, communication within and with environmental factors, habitat loss and even death [24], and plays important role in geographical distribution of these animals [25].

From all above discussion, it can be clearly seen that health effects of the noise are common as hypertension, myocardial infarction, stroke, mortality, dizziness, high blood pressure and cognitive diseases, irrespective of the noise sources, suggesting the control of all the noise sources [26]. To control the noise, different methods and equipment are made to control or minimize noise from different places like hospitals, educational institutions and workplaces [27]. Adaptive noise control, [28], active noise control, [29], shelter belts [30], equipment and home insulations [31] and active vibration noise control devices are made for this purpose and to reduce the risks of noise [32].

2. Sources of noise pollution

Noise pollution has many sources of which the traffic noise could be a major source. Other types include community noise, household, industrial, aircraft and ships noise [33–41].

2.1 Traffic vehicles

Traffic noise, as depicted by the name, is the noise originated from the traffic vehicles specially the old vehicles with no maintenance and those vehicles which have not been physically cleared to be driven on the roads. Heavy traffic vehicles also contribute in noise generation due to their heavy engines and load [42, 43]. Traffic noise represents the important environmental risk factors in mechanized areas [44] and it is one of the fastest growing and most ubiquitous types of environmental pollution [1]. It is associated with high blood pressure, but the long-term impact may lead to hospitalization and rare chances of death. In a study among London's 8.6 million residents, traffic noise was associated with cardiovascular defects in all adults (≥25 years) and the elderly (≥75 years). It shows that long-term exposure to road traffic noise increases the risk of death and the risk of cardiovascular disease in the population [45].

2.2 Commercial and Industrial noise

Different commercial activities like transportation of goods from one place to other using ships and heavy trucks create considerable noise in the respective areas. Ocean noise levels are increasing as a result of major growth in global trading activities which shows that if this activity continued to grow, which is by 1.9% each year, the contribution of commercial shipping to ambient ocean noise levels will be expected to intensely increase [46]. In addition to the ships, commercial aircrafts are also contributors in commercial and industrial noise. The main source of noise in aircrafts is the engine which generates more noise if the load on it is more [47, 48]. It is a well-known fact that all the machines produce noise and it is called as industrial noise. Different industries have different machinery like textile industry, wood industry and steel mills [49].

To assess the industrial noise effects on workers, all the workers of an industry in Jordan were included in a study. A structured questionnaire was used to collect data. Results revealed that out of total 191 workers, 145 (exposed to higher noise levels) had the issues of hypertension compared to those exposed to noise level lower than the permissible limit indicating that exposure to high level of noise is associated with elevated health risks [50].

2.3 Community noise

Community noise comprised of the noise during a match in the ground, traffic vehicle noise during waste collection, children playing in the streets, dogs barking, and noise during parties [51]. Musical instruments being played can also be a source of noise for someone not interested in music [52, 53]. One study assessed the noise and noise sensitivity of 364 adults living in the South African community and compared them with similar studies conducted in Switzerland. Compared to Swiss research, the proportion of people with high levels of noise sensitivity is higher (women: 35.1% vs. 26.9%, men: 25% vs. 20.5%), people are very angry about road traffic noise (women: 20, 5% compared to 12.4%; male: 17.9% vs. 11.1% were observed in South Africa). Although women in South Africa are more averse to community noise than Switzerland (21.1% vs. 9.4%), this is not the case for men (7.1% vs. 7.8%), suggesting that the noise pollution can seriously affect the inhabitants of the population noise [54].

2.4 Air craft noise

Airplanes such as army, navy and commercial aircraft are noise sources [55]. Airplanes, including army, naval and commercial aircraft, have become one of the most important environmental factors in terms of noise, and industry has identified much of their efforts and concerns. Noise abatement is the focus of modern research and development. Too much noise obviously damages our physical and mental health, so it makes sense to make a technical assessment of the noisy technology. Conflicts of interest in connection with aircraft noise are known. Propeller aircraft are the dominant noise. Many factors that contribute to the sound field in the propeller aircraft lead to extensive research to identify and improve the internal noise reduction techniques as noise sources [56].

2.5 Shipping noise

Anthropogenic noise is now considered a global problem, and recent research has shown that many animals have a multitude of negative effects. Marine

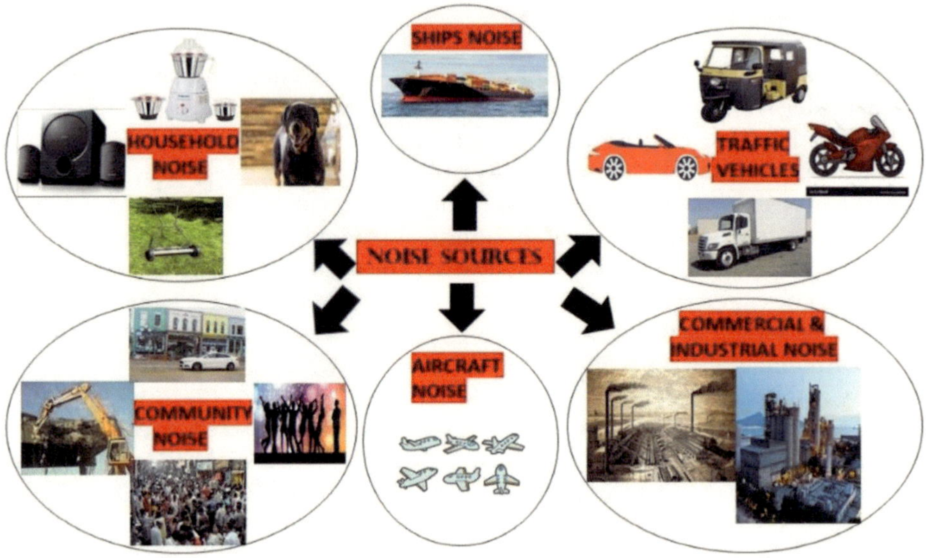

Figure 1.
Different noise sources.

underwater noise is increasingly considered an important and omnipresent pollutant that can affect marine ecosystems globally [57]. Ships and sea vessels are the source of this noise. Noise exposure varies markedly between the sites according to the number of the ships and sea vessels [58]. Due to the shipping noise, marine mammals and other marine life is vulnerable to be at risk as they require relatively quiet place to live but shipping noise could have substantial impacts on them and cause migration [59] (**Figure 1**).

3. Types of noise

Noise has different types according to the intensity, duration and frequency as continuous noise, intermittent noise, impulsive noise and low frequency noise.

3.1 Continuous noise

Continuous noise means the same noise frequency, intensity and quantity which is supplied to people and workers for longer periods of time like machinery operation in textile industry has the same amount, frequency and intensity for 6–8 hours of a working shift [60]. It is the noise which affects the industrial workers health badly by causing headaches to high blood pressure and other heart problems [61].

3.2 Intermittent noise

Noise which can occur at regular or irregular intervals is intermittent. This type of noise includes all the traffic vehicles noise and community noise as these are not the regular and continuously produced and varies with source [62–65].

3.3 Impulsive noise

This type of noise is produced instantly and reduced in the same way. Its types include ticking of clock, striking of hammer on something, water drops falling from

No.	Category/Area	In 1st July 2010		From 1st July 2013	
		Limits in (dB)			
		Day time	Night time	Day time	Night time
1	Residential area	65	50	55	45
2	Commercial area	70	60	65	55
3	Industrial area	80	75	75	65
4	Silence zone	55	45	50	45

Table 1.
Noise level limits in different categorized areas of the city (Pakistan).

height, and all the other noises in impulse forms. This type of noise also disturbs communication between people, induces stress, and anxiety in experimental population [66]. Mitigation of impulsive noise is extensively studied in wireline, wireless radio, and powerline communication systems [3, 67, 68].

3.4 Low frequency noise

Low frequency noise is common with background noise in urban environments and comes from road vehicles, aircraft, industrial machinery, artillery and mining explosions, wind turbines, compressors and ventilation or air conditioning systems. The effects of low-frequency noise are worrying because they are universal (effective propagation) compared to many structures, and many structures (home, wall and hearing protection) are less effective at attenuating low-frequency noise than other sounds. Intense low-frequency sounds seem to produce obvious symptoms, including respiratory disorders and hearing pain. Although it is difficult to determine the effect of low-frequency noise for methodological reasons, there are indications that some of the adverse effects of noise are usually due to low-frequency noise: loudness ratings and disturbing responses are sometimes given for equal sound pressure levels. Low frequency noises are greater than other noises, hum or vibration caused by low frequency noise amplify problems, and speech intelligibility can be reduced by low frequency noise more than other sounds, except for noise in the frequency range of the speech itself due to the upward propagation of the masking [69–72] (**Table 1**).

4. Health effects of noise

In light of the in-depth studies presented in **Table 2**, it can be concluded that traffic accounts for 80% of the environmental impact of noise [73]. It is generally believed that deafness, high blood pressure, ischemic heart disease, discomfort and insomnia, as well as effects on the immune system, are the cause of the noise pollution. In addition to the above diseases, headaches, dizziness, sleeplessness, high blood pressure and hypertension are the common diseases caused by noise.

4.1 On humans

Traffic noise emissions consist of complex components, including horns, engine noise and tire friction. It is estimated that the noise pollution is affected by traffic noise [74], learning disabilities, loss of communication and lack of attention [75, 76]. Epidemiological studies have shown that traffic noise increases the frequency of arterial diseases, hypertension and strokes as well as vascular dysfunctions [77]. Non-hearing effects such as activity, sleep and communication disorders can trigger a range of emotional reactions, including nuisance and subsequent stress, increased blood pressure and dyslipidemia, increased blood viscosity and blood sugar, and activation of the blood coagulation factor [78, 79]. Due to noise pollution, also higher memory disturbances and oxidative stress were observed [80].

4.1.1 On industrial workers

Occupational noise exposure to workers have adverse effects on workers' health by increasing hypertension, sleep disturbance [81], cardiovascular diseases, blood pressure, hypertension [82, 83], exhaustion and overworking, mistakes performed in various operations due to noise disturbance [84], memory impairment [85], increased pulse rate [86], hearing loss and diabetes [87, 88].

4.1.2 On public

Technology, modernization and residential complexes usually occur near the population, so that the resulting increase in noise is recorded. The environmental impact of noise is closely related to health consequences, including discomfort, sleep disorders and cardiovascular disease [89, 90]. In addition, noise can seriously damage communication, memory function and hearing [91].

4.2 On animals

Noise levels are steadily increasing worldwide and may potentially affect many animal species. Short-term exposure can affect the behavior and physiology of birds, reproductive system as birds avoid reproduction in noisy places [92].

Sr. no.	**Human health effects**		**Effects on animals**	
	Effects	**References**	**Effects**	**References**
1	Headache	[97, 98]	Hearing loss	[99]
2	Dizziness	[100]	Increased heart rate	[101]
3	Annoyance	[102]	Increased risk of death	[103]
4	High blood pressure	[104]	Habitat loss	[105]
5	Hypertension	[106]	Trouble in finding prey	[107]
6	Hearing loss	[108]	Trouble in finding mates as in frogs	[109]
7	Depression	[110]	Impaired reproduction in marine mammals	[111]
8	Sleeplessness	[112]	Affects balance system of squid	[113]
9	Physiological Stress	[114]		
10	Irritation	[115]		
11	Difficulty in communication	[116]		
12	Nervousness	[117]		

Table 2.
Noise induced health effects on humans and plants.

Animals also suffer human like disabilities like hearing loss, loss in responsiveness, dizziness and disturbance [93]. Traffic noise reduced foraging efficiency in most bats [94]. Monkeys also live in noise free areas as exhibited by a study in which continuous noise was supplied in the habitat of the monkeys in Brazil, monkeys moved from that area to noise free area indicating that they also do not like noise [95]. Noise effects on wildlife have also been widely studied and results indicated that they also prefer to live away from noise like bears, wolves, ants, lions and larger animals like elephants and whales [96] (**Table 2**).

5. Noise control and minimization

As it is widely discussed that noise has different negative effects on almost all the inhabitants of the planet, it is also one of the priorities that noise should be minimized to avoid the negative health impacts on the humans and animals. Due to its different sources, same technology cannot be used to address all types of noise. So, different technologies are adopted worldwide to overcome the noise impacts. Some of them are discussed below.

5.1 In industrial areas

The incorporation of sustainable industrial planning and development cannot be achieved without proper addressing of the noise pollution. Despite the proven impacts of noise pollution on the worker's health, there remains a lack of systematic methods to reduce the impacts of noise within the industries [118, 119]. One of the major advances have been recently demonstrated for the long-term noise minimizing technology from the Department of Defense, Australia to reduce the workplace exposure from military and industrial noise sources [120]. Application of personal protective equipment like ear plugs and ear mufflers are the general safety measures which are taken by the employer and workers themselves [121]. Sealing of the machinery by using rubber is also used to minimize the vibrations and noise of the machinery in some industries [122]. Glass industry is among the loudest noise producing industries and the workers in these industries are advised to use hearing protection equipment like ear plug or mufflers [123].

A recent technology called operating room technology is set up to address the occupational noise effects in which all the workers of the operating team are issued headsets with microphones. The headsets filter out background noise and the microphones enable interactive communication among and between [124].

5.2 In residential areas

Noise in residential areas is produced by traffic, celebration parties, loud music and playgrounds. Although the noise is a non-market good, the attempts of its evaluation have been increasing, usually by estimating the economic costs arising from exposure to noise, lost prices of property and medical expenses. It is estimated that plots located in the zone with noise exceedance the limits are 57% cheaper than those located in silent zone [125, 126]. The noise in the residential areas can be minimized by shifting to silent zones, using porous materials for house building or porous filters and planting hedges. In addition to this, some suggestion are also given by the authorities like exclusion of the traffic and train horn in or near residential areas, inclusion of noise barriers and removal of the railway tracks [127–129].

5.3 In commercial areas

There is an urgent need to reduce commercial noise being increased day by day due to increased commercial activities [130, 131]. Noise in these areas can be controlled or minimized by limiting transportation activities in markets in daytime, limiting commercial aircraft flights or changing the flight time to night [132].

5.4 In hospitals

Problems related to environmental noise are not confined outside the hospitals, but it became a major issue in the hospitals too. It needs to be solved because silence and peace in hospitals is a major contributor in healing of the patients [133]. Noise pollution in the operating rooms is one of the remaining challenges. Both patients and physicians are exposed to different sound levels during the operative cases, many of which can last for hours. For noise monitoring and control, sound sensors can be installed in patient bed spaces, hallways, and common areas to measure the noise levels and its control accordingly [134, 135]. Reactive noise barriers can also be installed in hospital facilities [136]. As we know that noise is produced from vibration, friction, collision and shocks, so by avoiding these phenomena's, we can avoid noise by using rubber and proper lubrication in machinery [137].

5.5 In educational institutions

The learning environment dramatically affects the learning outcomes of students. Noise is a major factor which can distract students [138], induce attention loss and concentration difficulties, anxiety and headache [139]. To address these problem in educational institutions, traffic noise should be regulated, and traffic can be banned accordingly if found exceeding. School and college bus-stands should be away from the school and colleges. Educational institutions should not be established near the railway tracks, stations and airports [140–142].

6. Conclusion and recommendations

There is a little difference between noise and a sound. A sound can be a noise when it is loud and intensive. Categorization of a sound and noise is also depending upon the choice of the listener and the circumstances. For example, rock music can be pleasurable sound to one person and could be an annoying noise to another person. Likewise, dog's barking is not a noise, but when it becomes continuous and disturbs people, it can be regarded as noise.

Noise has different sources and it can cause hearing loss, dizziness, heat diseases, headaches, high blood pressure, hypertension and nausea based upon the its intensity. To control environmental noise, different techniques and equipment are used according to the situation and requirement. In industries, ear plugs, and mufflers are used, rubber sealing and noise sensors are used in hospitals and shelter belts are used in residential areas for protection against adverse effects of noise. For future, it is necessary to work on reducing the vehicular sources of environmental noise by developing low noise producing automobile vehicles, aircrafts and ships.

Acknowledgements

The authors are highly acknowledged to Mr. Junaid Latif and Mr. Waqas Mohy Ud Din for their precious time given to review this manuscript.

Conflict of interest

The authors declared no conflict of interest for this chapter's publishing.

Author details

Zia Ur Rahman Farooqi*, Muhammad Sabir, Nukshab Zeeshan, Ghulam Murtaza, Muhammad Mahroz Hussain and Muhammad Usman Ghani
Institute of Soil and Environmental Sciences, University of Agriculture, Faisalabad, Pakistan

*Address all correspondence to: ziaa2600@gmail.com

References

[1] Zhao J et al. Assessment and improvement of a highway traffic noise prediction model with L eq (20s) as the basic vehicular noise. Applied Acoustics. 2015;**97**:78-83

[2] Nakashima H, Shinkai T, Kakinuma T. Vehicular suction noise transmission system. Google Patents; 2018

[3] Liu S et al. Double kill: Compressive-sensing-based narrow-band interference and impulsive noise mitigation for vehicular communications. IEEE Transactions on Vehicular Technology. 2016;**65**(7):5099-5109

[4] Garg N et al. Effect of odd-even vehicular restrictions on ambient noise levels in Delhi city. In: 2017 International Conference on Advances in Mechanical, Industrial, Automation and Management Systems (AMIAMS); 2017

[5] Cai M et al. Road traffic noise mapping in Guangzhou using GIS and GPS. Applied Acoustics. 2015;**87**:94-102

[6] NIOSH. Preventing Occupational Hearing Loss—A Practical Guide. Cincinnati, Ohio: DHHS (NIOSH) Publication; 1996. pp. 96-110

[7] Klein A et al. Spectral and modulation indices for annoyance-relevant features of urban road single-vehicle pass-by noises. The Journal of the Acoustical Society of America. 2015;**137**(3):1238-1250

[8] Cole JA, Luthey-Schulten Z. Careful accounting of extrinsic noise in protein expression reveals correlations among its sources. Physical Review E. 2017;**95**(6):062418

[9] Beimer W, Maennig W. Noise effects and real estate prices: A simultaneous analysis of different noise sources. Transportation Research Part D: Transport and Environment. 2017;**54**:282-286

[10] Geraghty D, O'Mahony M. Urban noise analysis using multinomial logistic regression. Journal of Transportation Engineering. 2016;**142**(6):04016020

[11] Koruk H, Arisoy A. Identification of crack noises in household refrigerators. Applied Acoustics. 2015;**89**:234-243

[12] Ismail Y, Shadid N, Nizam A. Development of green curtain noise barrier using natural waste fibres. Journal of Advanced Research in Materials Science. 2016;**17**(1):1-9

[13] Sun K, Neitzel RL. What can 35 years and over 700,000 measurements tell us about noise exposure in the mining industry? AU - Roberts, Benjamin. International Journal of Audiology. 2017;**56**(Suppl 1):4-12

[14] Wang Z et al. Noise hazard and hearing loss in workers in automotive component manufacturing industry in Guangzhou, China. Zhonghua Lao Dong Wei Sheng Zhi ye Bing za Zhi [Chinese Journal of Industrial Hygiene and Occupational Diseases]. 2015;**33**(12):906-909

[15] Rizkya I et al. Measurement of noise level in enumeration station in rubber industry. IOP Conference Series: Materials Science and Engineering. 2017;**180**:012121

[16] Nassur A-M et al. The impact of aircraft noise exposure on objective parameters of sleep quality: Results of the DEBATS study in France. Sleep Medicine. 2019;**54**:70-77

[17] Schmidt FP et al. Crucial role for Nox2 and sleep deprivation in aircraft noise-induced vascular and cerebral oxidative stress, inflammation, and gene

regulation. European Heart Journal. 2018;**39**(38):3528-3539

[18] Lawton RN, Fujiwara D. Living with aircraft noise: Airport proximity, aviation noise and subjective wellbeing in England. Transportation Research Part D: Transport and Environment. 2016;**42**:104-118

[19] Badino A et al. Airborne noise emissions from ships: Experimental characterization of the source and propagation over land. Applied Acoustics. 2016;**104**:158-171

[20] Van Kamp I et al. Burden of disease from exposure to low frequency noise: A Dutch inventory. In: ICBEN Proceedings; 2017

[21] Michaud DS et al. Exposure to wind turbine noise: Perceptual responses and reported health effects. The Journal of the Acoustical Society of America. 2016;**139**(3):1443-1454

[22] Recio A et al. Road traffic noise effects on cardiovascular, respiratory, and metabolic health: An integrative model of biological mechanisms. Environmental Research. 2016;**146**:359-370

[23] Baliatsas C et al. Health effects from low-frequency noise and infrasound in the general population: Is it time to listen? A systematic review of observational studies. Science of the Total Environment. 2016;**557-558**:163-169

[24] Tennessen JB, Parks SE, Langkilde TL. Anthropogenic Noise and Physiological Stress in Wildlife. New York, NY: Springer; 2016

[25] Shieh B-S et al. Interspecific comparison of traffic noise effects on dove coo transmission in urban environments. Scientific Reports. 2016;**6**:32519

[26] Stansfeld SA. Noise effects on health in the context of air pollution exposure. International Journal of Environmental Research and Public Health. 2015;**12**(10):12735

[27] Ferrer M et al. Active noise control over adaptive distributed networks. Signal Processing. 2015;**107**:82-95

[28] Wurm M. Adaptive noise control. Google Patents; 2015

[29] Christoph M, Wurm M. Active noise control system. Google Patents; 2016

[30] Dhillon GS et al. Spectroscopic investigation of soil organic matter composition for shelterbelt agroforestry systems. Geoderma. 2017;**298**:1-13

[31] Van Renterghem T. Green roofs for acoustic insulation and noise reduction. In: Nature Based Strategies for Urban and Building Sustainability. United Kingdom: Butterworth-Heinemann-Elsevier; 2018. pp. 167-179

[32] Ohta Y et al. Active vibration noise control device. Google Patents; 2015

[33] Abbaspour M et al. Hierarchal assessment of noise pollution in urban areas—A case study. Transportation Research Part D: Transport and Environment. 2015;**34**:95-103

[34] Aso N et al. Moulins detected as ambient noise sources at the Kaskawulsh Glacier. In: AGU Fall Meeting Abstracts; 2016

[35] Droitcour AD et al. Sources of noise and signal-to-noise ratio. In: Doppler Radar Physiological Sensing. New Jersey: John Willey and Sons; 2016. pp. 137-170

[36] Zannin PHT et al. Evaluation of environmental noise generated by household waste collection trucks. Journal of Environmental Assessment Policy and Management. 2018;**20**(04):1850010

[37] Antoniali M, Versolatto F, Tonello AM. An experimental characterization of the PLC noise at the source. IEEE Transactions on Power Delivery. 2016;**31**(3):1068-1075

[38] Dale LM et al. Socioeconomic status and environmental noise exposure in Montreal, Canada. BMC Public Health. 2015;**15**(1):205

[39] Li S-Q, Xiao W, Sun X-X. Household appliance fault detection based on wavelet denoising and HHT. In: DEStech Transactions on Computer Science and Engineering; 2018

[40] Yang W, Moon HJ. Effects of recorded water sounds on intrusive traffic noise perception under three indoor temperatures. Applied Acoustics. 2019;**145**:234-244

[41] Clark C et al. Association of long-term exposure to transportation noise and traffic-related air pollution with the incidence of diabetes: A prospective cohort study. Environmental Health Perspectives. 2017;**125**(8):087025-087025

[42] Wang X et al. Development of a road shoulder's equivalent sound source traffic noise prediction model. Proceedings of the Institution of Civil Engineers Transport. 2019;**0**(0):1-10

[43] Ware HE et al. A phantom road experiment reveals traffic noise is an invisible source of habitat degradation. Proceedings of the National Academy of Sciences. 2015;**112**(39):12105-12109

[44] Schmidt FP et al. Environmental stressors and cardio-metabolic disease: Part I—Epidemiologic evidence supporting a role for noise and air pollution and effects of mitigation strategies. European Heart Journal. 2017;**38**(8):550-556

[45] Halonen JI et al. Road traffic noise is associated with increased cardiovascular morbidity and mortality and all-cause mortality in London. European Heart Journal. 2015;**36**(39):2653-2661

[46] Kaplan MB, Solomon S. A coming boom in commercial shipping? The potential for rapid growth of noise from commercial ships by 2030. Marine Policy. 2016;**73**:119-121

[47] Hu Y et al. Commercial aircraft cabin noise reduction based on SEA and transfer-matrix method. Noise Control Engineering Journal. 2018;**66**(4):362-374

[48] Meng Q, Sun Y, Kang J. Effect of temporary open-air markets on the sound environment and acoustic perception based on the crowd density characteristics. Science of the Total Environment. 2017;**601-602**:1488-1495

[49] Bozkurt TS, Demirkale SY. The field study and numerical simulation of industrial noise mapping. Journal of Building Engineering. 2017;**9**:60-75

[50] Nserat S et al. Blood pressure of Jordanian workers chronically exposed to noise in industrial plants. International Journal of Occupational and Environmental Medicine. 2017:**8**(4):217-223

[51] Weuve J et al. Long-term exposure to community noise in relation to Alzheimer's disease and related dementias. In: ISEE Conference Abstracts; 2018

[52] Slater J et al. Music training improves speech-in-noise perception: Longitudinal evidence from a community-based music program. Behavioural Brain Research. 2015;**291**:244-252

[53] Keith S, Michaud D. Estimating community tolerance for wind turbine noise annoyance. The Journal of the Acoustical Society of America. 2017;**141**(5):3727-3727

[54] Sieber C et al. Comparison of sensitivity and annoyance to road traffic and community noise between a South African and a Swiss population sample. Environmental Pollution. 2018;**241**:1056-1062

[55] Khozikov V et al. Synergetic noise absorption and anti-icing for aircrafts. Google Patents; 2017

[56] Ommi F, Azimi M. Low-frequency interior noise in prop-driven aircrafts: Sources and control methodologies. Noise & Vibration Worldwide. 2017;**48**(7-8):94-98

[57] Williams R et al. Impacts of anthropogenic noise on marine life: Publication patterns, new discoveries, and future directions in research and management. Ocean and Coastal Management. 2015;**115**:17-24

[58] Merchant ND et al. Soundscape and Noise Exposure Monitoring in a Marine Protected Area Using Shipping Data and Time-Lapse Footage. New York, NY: Springer; 2016

[59] Halliday WD et al. Potential impacts of shipping noise on marine mammals in the western Canadian Arctic. Marine Pollution Bulletin. 2017;**123**(1):73-82

[60] Han J-H, Dimitrijevic A. Continuous noise maskers reduce cochlear implant related artifacts during electrophysiological testing. The Journal of Health Science. 2017;**7**(2):61-62

[61] Usenko VC, Filip R. Trusted noise in continuous-variable quantum key distribution: A threat and a defense. Entropy. 2016;**18**(1):20

[62] Nichols TA, Anderson TW, Širović A. Intermittent noise induces physiological stress in a coastal marine fish. PLoS One. 2015;**10**(9):e0139157

[63] Khoshnevisan D, Kim K. Nonlinear noise excitation of intermittent stochastic PDEs and the topology of LCA groups. The Annals of Probability. 2015;**43**(4):1944-1991

[64] Biao W et al. An adaptive data detection algorithm based on intermittent chaos with strong noise background. Neural Computing and Applications. 2018;**0**:1-8

[65] Zhou H, Li W. Stabilisation of stochastic delayed systems with Lévy noise on networks via periodically intermittent control. International Journal of Control. 2018:1-14

[66] Radford AN et al. Repeated exposure reduces the response to impulsive noise in European seabass. Global Change Biology. 2016;**22**(10):3349-3360

[67] Kuai X et al. Impulsive noise mitigation in underwater acoustic OFDM systems. IEEE Transactions on Vehicular Technology. 2016;**65**(10):8190-8202

[68] Ma W et al. Sparse least mean p-power algorithms for channel estimation in the presence of impulsive noise. Signal, Image and Video Processing. 2016;**10**(3):503-510

[69] Silva MBD et al. A physics-based statistical RTN model for the low frequency noise in MOSFETs. IEEE Transactions on Electron Devices. 2016;**63**(9):3683-3692

[70] Fogarty MA et al. Nonexponential fidelity decay in randomized benchmarking with low-frequency noise. Physical Review A. 2015;**92**(2): 022326

[71] Van Kamp I. Determinants of Annoyance from Humming Sound as Indicator of Low Frequency Noise. Acoustics. 2017;**0**:1-7

[72] Li H et al. A method for low-frequency noise suppression based on mathematical morphology in

microseismic monitoring. Geophysics. 2016;**81**(3):V159-V167

[73] Tobías A et al. Health impact assessment of traffic noise in Madrid (Spain). Environmental Research. 2015;**137**:136-140

[74] Franklin M et al. Assessing the contribution of noise to the association between traffic-related air pollution and children's respiratory health. In: ISEE Conference Abstracts; 2018

[75] Connolly D et al. The effects of classroom noise on the reading comprehension of adolescents. The Journal of the Acoustical Society of America. 2019;**145**(1):372-381

[76] Mayor S. Noise Pollution: WHO Sets Limits on Exposure to Minimise Adverse Health Effects. London, UK: British Medical Journal Publishing Group; 2018

[77] Ullmann E et al. Effects of noise on vascular function, oxidative stress, and inflammation: Mechanistic insight from studies in mice. European Heart Journal. 2017;**38**(37):2838-2849

[78] Münzel T et al. The adverse effects of environmental noise exposure on oxidative stress and cardiovascular risk. Antioxidants & Redox Signaling. 2018;**28**(9):873-908

[79] Injaian AS, Taff CC, Patricelli GL. Experimental anthropogenic noise impacts avian parental behaviour, nestling growth and nestling oxidative stress. Animal Behaviour. 2018;**136**:31-39

[80] Sikandaner HE et al. Neuroprotective effects of sildenafil against oxidative stress and memory dysfunction in mice exposed to noise stress. Behavioural Brain Research. 2017;**319**:37-47

[81] Zare S et al. Evaluation of the effects of occupational noise exposure on serum aldosterone and potassium among industrial workers. Noise & Health. 2016;**18**(80):1-6

[82] Assunta C et al. Noise and cardiovascular effects in workers of the sanitary fixtures industry. International Journal of Hygiene and Environmental Health. 2015;**218**(1):163-168

[83] Lie A et al. Systematic review of the cardiovascular effects of occupational noise. Occupational Medicine. 2016; **66**(1):10-16

[84] Simion S et al. Estimation, Assessment and Effects of Workers Exposure to Physical Noxae. Cham: Springer International Publishing; 2018

[85] Burgess M, Molesworth B. Effects of noise for workers in the transportation industry. The Journal of the Acoustical Society of America. 2017;**141**(5):3804-3804

[86] Yousif KM, Ali FM. Effects of occupational noise exposure on blood pressure and other aspects of health for workers in selected industrial places in Zakho City, Iraq. Science Journal of University of Zakho. 2017;**5**(1):136-141

[87] Lie A et al. Occupational noise exposure and hearing: A systematic review. International Archives of Occupational and Environmental Health. 2016;**89**(3):351-372

[88] Cantley LF et al. Association between ambient noise exposure, hearing acuity, and risk of acute occupational injury. Scandinavian Journal of Work, Environment & Health. 2015;**41**(1):75-83

[89] Hays J, McCawley M, Shonkoff SBC. Public health implications of environmental noise associated with unconventional oil and gas development. Science of the Total Environment. 2017;**580**:448-456

[90] Lusk SL et al. Position statement: Harmful effects of environmental noise exposures. Nursing Outlook. 2016;**64**(4):395-396

[91] Zhidkov SV. Statistical characterization and modeling of noise effects in near-ultrasound aerial acoustic communications. The Journal of the Acoustical Society of America. 2018;**144**(4):2605-2612

[92] Halfwerk W, Slabbekoorn H, Both C. Noise affects nest-box choice of 2 competing songbird species, but not their reproduction. Behavioral Ecology. 2016;**27**(6):1592-1600

[93] Lai J, Bartlett EL. Masking differentially affects envelope-following responses in young and aged animals. Neuroscience. 2018;**386**:150-165

[94] Luo J, Siemers BM, Koselj K. How anthropogenic noise affects foraging. Global Change Biology. 2015;**21**(9):3278-3289

[95] Duarte MHL et al. Mining noise affects loud call structures and emission patterns of wild black-fronted Titi monkeys. Primates. 2018;**59**(1):89-97

[96] Meillere A et al. Traffic noise exposure affects telomere length in nestling house sparrows. Biology Letters. 2015;**11**(9):20150559

[97] Lambert GA, Zagami AS. Does somatostatin have a role to play in migraine headache? Neuropeptides. 2018

[98] Kim J et al. The relationship between occupational noise and vibration exposure and headache/eyestrain, based on the fourth Korean Working Condition Survey (KWCS). PLoS One. 2017;**12**(5):e0177846

[99] Sha S-H, Schacht J. Emerging therapeutic interventions against noise-induced hearing loss. Expert Opinion on Investigational Drugs. 2017;**26**(1):85-96

[100] Elsheshtawy M, Yang F, Prabhu S. Dizziness spells: Should one suspect the pacemaker? The American Journal of Emergency Medicine. 2018

[101] Lyamin OI et al. Cardiorespiratory responses to acoustic noise in belugas. In: The Effects of Noise on Aquatic Life II. Switzerland: Springer Nature; 2016. pp. 665-672

[102] Michaud DS et al. Personal and situational variables associated with wind turbine noise annoyance. The Journal of the Acoustical Society of America. 2016;**139**(3):1455-1466

[103] Da Fonseca J et al. Noise-induced duodenal lesions. European Journal of Anatomy. 2019;**9**(1):29-33

[104] Schmidt F et al. Nighttime aircraft noise impairs endothelial function and increases blood pressure in patients with or at high risk for coronary artery disease. Clinical Research in Cardiology. 2015;**104**(1):23-30

[105] Dähne M et al. Bubble curtains attenuate noise from offshore wind farm construction and reduce temporary habitat loss for harbour porpoises. Marine Ecology Progress Series. 2017;**580**:221-237

[106] Fuks KB et al. Long-term exposure to ambient air pollution and traffic noise and incident hypertension in seven cohorts of the European study of cohorts for air pollution effects (ESCAPE). European Heart Journal. 2017;**38**(13):983-990

[107] McCormick MI et al. Effect of elevated CO_2 and small boat noise on the kinematics of predator–prey interactions. Proceedings of the Royal Society B: Biological Sciences. 2018;**285**(1875):20172650

[108] Kujawa SG, Liberman MC. Synaptopathy in the noise-exposed and aging cochlea: Primary neural degeneration in acquired sensorineural hearing loss. Hearing Research. 2015;**330**:191-199

[109] Ferrari MC et al. School is out on noisy reefs: The effect of boat noise on predator learning and survival of juvenile coral reef fishes. Proceedings of the Royal Society B: Biological Sciences. 2018;**285**(1871):20180033

[110] Seidler A et al. Association between aircraft, road and railway traffic noise and depression in a large case-control study based on secondary data. Environmental Research. 2017;**152**:263-271

[111] Nabi G et al. The possible effects of anthropogenic acoustic pollution on marine mammals' reproduction: An emerging threat to animal extinction. Environmental Science and Pollution Research. 2018;**25**(20):1-8

[112] Kageyama T. Adverse effects of community noise as a public health issue. Sleep and Biological Rhythms. 2016;**14**(3):223-229

[113] Kim M et al. Regulatory evolution and voltage-gated ion channel expression in squid axon: Selection-mutation balance and fitness cliffs. PLoS One. 2015;**10**(4):e0120785

[114] Aminuddin R, Sharkey A. Does interacting with a robot reduce your physiological stress response. New Friends. 2016;**0**(0):1-2

[115] Schneider S et al. Structure-borne noise based diesel engine control. MTZ Worldwide. 2017;**78**(11):76-81

[116] Le Prell CG, Clavier OH. Effects of noise on speech recognition: Challenges for communication by service members. Hearing Research. 2017;**349**:76-89

[117] Geravandi S et al. Noise pollution and health effects. Jundishapur Journal of Health Sciences. 2015;**7**(1):e60312

[118] Hammad A, Akbarnezhad A, Rey D. Accounting for noise pollution in planning of smart cities. In: Smart Cities as a Solution for Reducing Urban Waste and Pollution. IGI Global; 2016. pp. 149-196

[119] Nasir MS et al. Evaluation and analysis of traffic noise in different zones of Faisalabad—An industrial city of Pakistan. Geology, Ecology, and Landscapes. 2017;**1**(4):232-240

[120] Teague P, Conomos J, Jennings M. Noise reduction advances at major defence facilities in Australia. In: INTER-NOISE and NOISE-CON Congress and Conference Proceedings. Institute of Noise Control Engineering; 2016

[121] Nawi NM et al. Occupational Noise Exposure of Construction Workers at Construction Sites in Malaysia. Singapore: Springer; 2018

[122] Hammad AWA, Akbarnezhad A, Rey D. A multi-objective mixed integer nonlinear programming model for construction site layout planning to minimise noise pollution and transport costs. Automation in Construction. 2016;**61**:73-85

[123] Lim MH et al. Strategic noise mapping prediction for a rubber manufacturing factory in Malaysia. In: E3S Web of Conferences. EDP Sciences; 2018

[124] Friedrich MG et al. New technical solution to minimise noise exposure for surgical staff: The 'silent operating theatre optimisation system'. BMJ Innovations. 2017;**3**(4):196-205

[125] Łowicki D, Piotrowska S. Monetary valuation of road noise. Residential property prices as an indicator of the

acoustic climate quality. Ecological Indicators. 2015;**52**:472-479

[126] Hao Y et al. On the relationship between traffic noise resistance and urban morphology in low-density residential areas. Acta Acustica united with Acustica. 2015;**101**(3):510-519

[127] Morandi F et al. Standardised acoustic characterisation of sonic crystals noise barriers: Sound insulation and reflection properties. Applied Acoustics. 2016;**114**:294-306

[128] Bunn F, Zannin PHT. Assessment of railway noise in an urban setting. Applied Acoustics. 2016;**104**:16-23

[129] Van Renterghem T et al. Using natural means to reduce surface transport noise during propagation outdoors. Applied Acoustics. 2015;**92**:86-101

[130] Williams R et al. Approaches to reduce noise from ships operating in important killer whale habitats. Marine Pollution Bulletin. 2019;**139**:459-469

[131] Fang Y, Huang J. Resolving weak light of sub-picowatt per square centimeter by hybrid perovskite photodetectors enabled by noise reduction. Advanced Materials. 2015;**27**(17):2804-2810

[132] Nelson JP. Airport noise, location rent, and the market for residential amenities. Journal of Environmental Economics and Management. 1979;**6**(4):320-331

[133] Gallo LF, Olivera J. Noise pollution control system in the hospital environment. Journal of Physics: Conference Series. 2016;**705**:012942

[134] Kawai Y et al. Quality improvement initiative to reduce pediatric intensive care unit noise pollution with the use of a pediatric delirium bundle. Journal of Intensive Care Medicine. 2019;**34**(5):383-390

[135] Fiedler PEK, Zannin PHT. Evaluation of noise pollution in urban traffic hubs—Noise maps and measurements. Environmental Impact Assessment Review. 2015;**51**:1-9

[136] Monazzam M, Babamahmoodi A, Naderzadeh M. Reactive noise barrier optimization plan for noise pollution reduction in fifteen districts of Tehran. Journal of Air Pollution and Health. 2016;**1**(1):27-34

[137] Korka ZI et al. Gearboxes noise reduction by applying a fluoropolymer coating procedure. Environmental Engineering and Management Journal. 2015;**14**(6):1433-1439

[138] Gilavand A, Jamshidnezhad A. The effect of noise in educational institutions on learning and academic achievement of elementary students in Ahvaz, South-West of Iran. International Journal Of Pediatrics. 2016;**4**(3):1453-1463

[139] Bitar ML, Calaço Sobrinho LF, Simões-Zenari M. Ações para a melhoria do conforto acústico em instituições de educação infantil. Ciência & Saúde Coletiva. 2018;**23**:315-324

[140] Gupta A et al. Noise pollution and impact on children health. The Indian Journal of Pediatrics. 2018;**85**(4):300-306

[141] Yucel M, Kahveci B, Colakkadioglu D. Modelling the adverse health effects of road traffic noise: A case study in Adana, Bulent Angin Boulevard. Journal of International Environmental Application & Science. 2017;**12**(4):325-333

[142] Yadav S, Bilas R. Effect of noise pollution on exposed people of Gorakhpur City. Indian Journal of Scientific Research. 2017;**8**(1):99-107

Chapter 2

Autonomous Vehicles toward a Revolution in Collective Transport

Sylvie Mira Bonnardel, Fabio Antonialli and Danielle Attias

Abstract

Autonomous public transport is not a fantasy anymore. At least 92 experimentations with autonomous shuttles for collective transport (ASCTs) have been running in different cities around the world. The deployment of autonomous shuttles opens up a wide range of possibilities in rethinking urban mobility, with a greater focus on users, new services, and reduced urban traffic congestion. The prospective analysis of these possibilities produces original, innovative forecasts with new scenarios for ASPT based on three strong drivers of change, i.e., technological innovation, new mobility behavior, and new business models for public transport. Furthermore, the deployment of ASCTs will considerably modify costs of urban externalities and extend inclusion. The chapter presents how the implementation of autonomous shuttles may revolutionize the public transport.

Keywords: autonomous shuttle, prospective, public transport transformation, autonomous shared mobility

1. Introduction

With 55% of the world's current population living in urban areas and estimates of 68% by 2050, mobility has been playing a fundamental role on our daily lives. It not only affects our well-being and life quality, but it also dictates where people work, were people live, and thereby on how people commute [1–3].

Nevertheless, our current means of transport are reaching their social, economic, and environmental limits. Our car-dependent and fossil-fuel-reliant societies are becoming sources of major problems, such as increased congestion, severe air pollution, noise, and many other externalities associated with transporting people and goods. Thereby, there has been a worldwide consensus that efforts are needed to develop more sustainable transport solutions [4].

Recent advances and innovations in technology and digitalization are presenting significant impacts on designing sustainable mobility concepts to counteract this trend [5]. The complexity of the automotive sector has dramatically grown over the past 10 years, mainly due to technology disruption, changes in regulations, as well as transformations in users' behavior [6].

As highlighted by many authors, on-demand mobility services, autonomous driving, dynamic pricing algorithms, and vehicle electrification will change the way people experience mobility in urban environments. This is clearly highlighted by a 2018 report from PriceWaterhouseCoopers [7]; according to the company's report, the car of the future will be EASCY: electrified, autonomous, shared, connected, and yearly updated.

That is, the mobility of the future will be much easier, more flexible, and user-centered. In this new mobility paradigm, autonomous vehicles (AVs) are expected to play an important role in urban mobility contexts, reducing the number of accidents and pollution, reducing transport costs and time, improving traffic efficiency as well as productivity, and promoting social inclusion for those who cannot or do not want to drive [8–10]. However, the proliferation of AVs is far from guaranteed, since complex issues related to legal aspects, liability, privacy, licensing, cyber-security, insurances, and so on still need to be further addressed and developed [11, 12].

Thereby fleets of private autonomous cars are not likely to be seen on roads and streets in a perceivable feature [13]. The most revolutionary impact of AVs will probably be on collective public transports. The introduction of on-demand mobility in collective public transport will deeply transform collective transport uses and business models, thereby undoubtedly attending to a real paradigm shift [14].

Autonomous shuttles for collective transport (ASCTs) are a technological urban mobility solution aimed at meeting the demand of first- and last-mile commute as well as microtransit. Their emergence promises to take advantage of connected AVs to allow Mobility-as-a-Service (MaaS) schemes [15], with a wide array of scenarios' deployments, such as urban centers, central commercial districts, university campuses, airports, shopping malls, hospitals, etc.

Within this context, a significant group of entrants have been carrying out trials and demonstrations worldwide with ASCTs [16]. However, like many other technology push innovations, there is still much to learn about the operation of such vehicles, not only from policy and regulation perspectives but also regarding business models, impact on communities, user acceptance, and demand [17].

In this sense, this chapter proposes to discuss the most relevant experimentations with ASCTs worldwide, by presenting an overview of the countries and cities with most deployments, the most prominent shuttle manufacturers, and the most recurrent typologies of use, prevailing business models, and so on. It also aims to provide an in-depth view on selected successful deployments of ASCTs, especially regarding their social and economic impacts, and at last, we intend to lay grounds for what will be the future social, economic, and environmental implications of the insertion of ASCTs. We intend to lay grounds on a new paradigm for collective transport, via shared on-demand mobility.

This study is part of the European project AVENUE, the Autonomous Vehicles to Evolve to a New Urban Experience project [18], an EU funded project which debuted on May 1, 2018, and will last for 48 months. It aims to design and carry out full-scale demonstrations of urban transport automation by deploying, fleets of autonomous shuttles on mixed traffic conditions in low-to-medium-demand areas of four European demonstrator cities: Geneva, Lyon, Copenhagen, and Luxembourg, by providing innovative services, like on-demand door-to-door transportation and multimodal commute solutions.

Besides this introduction, the chapter is structured as follows: Section 2 presents an overall benchmark on 92 experimentations with ASCTs worldwide. Section 3 presents a more in-depth analysis on 11 of these 92 experimentations, divided into three main European projects. At last on Section 4, we propose a prospective view of ASCTs implementation that pictures how it may revolutionize collective transport.

2. Worldwide benchmark with ASCTs

The implementation of ASCTs can be considered as a building block for the urban mobility revolution, in a sense that it involves a wide range of stakeholders from the most diverse fields as well as it pressures policy makers to act on changing and updating regulations [19].

In fact, as pointed out by Harris (2018), there is still much to learn about the operation of ASCTs from policy and regulation perspective as well as regarding business models and consumer acceptance. However, information regarding the scope of the implementations is still scarce, nonstructured, and pulverized.

Thereby, Antonialli [20] carried out an international benchmark study of ASCTs as a way to better understand such issues. The author's research (carried out on September 2018) was not limited to current (ongoing) experimentations; in this sense, finished projects as well as projects yet-to-start were also considered. Over the next pages, we briefly summarize the author's main findings.

In total, 92 experimentations were sampled, of which 50 had already been finished, 31 were currently running, and 11 were still going to be initiated. These 92 projects unfold in 78 cities spread over 32 countries worldwide.

Figure 1 depicts the geographical dispersion of the projects worldwide. Europe is on the lead regarding the total number of experimentations, that is, from the 32 countries that composed the sample, the continent holds 20 and that together comprise a total of 53 from the 92 projects (58%). Next comes Asia with 15 experimentations across 8 countries (16%), followed by North America (14%) with 13 projects divided among the USA and Canada, and, at last, Oceania presents 11 projects spread across Australia and New Zealand (12%). It is worth noting that no country in South America, Central America, and Africa were present in the sample.

Another interesting analysis is that Europe is not only on the lead when it comes to the total number of experimentations, the continent also holds the largest number of shuttle manufacturers. According to the author, Europe holds 9 of the total of 20 OEMs, which are responsible for providing the shuttles for 80 out of the 92 experimentations (a total of 87%).

Notwithstanding, it is important to highlight the relevance of the French startups Navya and EasyMile. The two companies are global leaders regarding both manufacturing and deployments of autonomous shuttles.

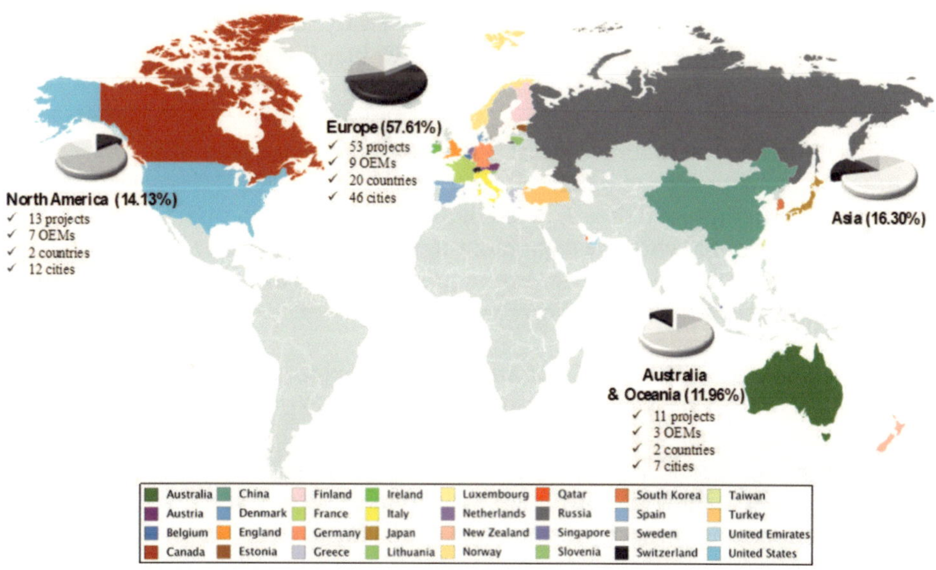

Figure 1.
Experimentations with ASCTs worldwide. Source: [20].

	EZ10 by EasyMile	ARMA by Navya
Capacity:	12 passengers (6 sitting and 6 standing)	15 passengers (8 sitting and 7 standing)
Cruising speed:	20 km/h	25 km/h
Maximum speed:	40 km/h	45 km/h
Propulsion engine:	Electric	Electric
Length:	3,93 meters	4.75 meters
Width:	1,99 meters	2.05 meters
Height:	2,75 meters	2.55 meters
Vehicle cost:	200,000 to 220,000€. (\$223,180 to \$245,498)	200,000€ (\$223,180)
Maintenance costs:	30,000€/year (\$33,477/year)	90,000€/year (\$101,511/year)

Figure 2.
Navya's ARMA and EasyMile's EZ10 technical specifications. Source: [20].

With the greatest number of ongoing projects and projects yet-to-start (16 and 2, respectively), Navya is the current market leader. The company was founded in 2014 in Lyon and launched their ARMA shuttle (with capacity of 15 passengers) in October 2015. On November 2017, they launched a new product, named "Autonom Cab," which they claim to be the first robot-taxi in the market. The vehicle, with capacity for six passengers, was designed to work as an on-demand service, for both hide-hailing and shared hide-hailing (which would be autonomous counterparts of services like Uber and UberPool).

EasyMile, on the other hand, is seen as Navya's main competitor [21]. The company is the result of a joint venture between Ligier (vehicle manufacturer) and Robosoft (high-tech robotics company). It was also founded in 2014 with its headquarters in Toulouse, and their autonomous shuttle, the EZ10 (with capacity for 12 passengers), was developed with the help of the CityMobil2 project, and it has been deployed on projects in Australia, Canada, China, Estonia, Finland, France, Germany, Ireland, Japan, the Netherlands, Norway, Singapore, Spain, Sweden, Switzerland, the United Arab Emirates, and the United States.

Figure 2 depicts the main technical features of both shuttles.

Autonomous shuttles can be described as a "technology push" innovation. In this sense, in order for ASCTs to succeed in the market, it is essential for the general public to be acquainted with the technology and its use forms, not only to cease their inherent human curiosity but also as a way to build trust.

In this sense, the benchmark shows a predominance of:

- Showcases (21%): Where ASCTs are demonstrated to potential consumers in hopes of (1) getting them acquainted to it and/or (2) getting them to acquire it

- Trials (69%): Temporary offering (mainly in fixed, looped routes in controlled environments) allowing consumers to examine, use, or test the ASCT prior to fully committing company resources to a full launch

- Regular services (10%): permanent (paid) transportation service from point A to B via an ASCT

Thereby, due to this inherent "technology push" nature of the deployments, most of the experimentations were mainly offered free of charge to riders (94% of the total sample). As for the road environment, 52% of the deployments were set to take place in closed/controlled areas (e.g., university campuses, parks, hospitals, resorts, airports, and other designated roads); the remaining 48% were set to run in mixed traffic conditions with low-to-medium-demand areas.

Since the testing with autonomous vehicles is not yet legal in all countries and regions, it is worth highlighting that in all 92 sampled deployments, a human operator was required to be onboard the vehicle at all times as well as the operating speeds were not higher than 15 km/h.

91% of the experimentations were classified as regular-line transport (RLT). On the other hand, only 4.5% of projects were fit under demand-responsive transport (DRT), and the last remaining 4.5% were offered in both RLT and DRT modes. Thereby, as stated by the author, as more countries and cities begin to allow testing the circulation of AVs, the percentage of DRT autonomous mobility is likely to increase, since the major value proposition claimed by ASCTs' manufacturers is to facilitate the first- and last-mile commute as well as microtransit.

Based on the aforementioned elements, a set of typologies of use for ASCTs can be synthetized in **Table 1**.

Regarding the business models toward private transportation (in blue), only three experimentations were found, all within the scope of regular-line transport, those being: May Mobility (American shuttle manufacturer) transporting Quick Loans workers in Detroit from parking lots to their office buildings; (typology 2) and Navya transporting workers in the Civaux nuclear power plant in France and in Sydney Olympic Park in Australia (typology 3).

Regarding public transportation offerings, a total of 89 experimentations were identified, covering all 8 of the typologies proposed. 79 of these experiments (89%) fit RLT, with 23 (26%) projects in typology 1 (first- and last-mile RLT in closed traffic), 17 (19%) in typology 2 (first- and last-mile RLT in mixed traffic), 18 (20%) in typology 3 (microtransit RLT in closed traffic), and 21 (24%) within typology 4 (microtransit RLT in mixed traffic). The remaining projects (11%) in public transportation contexts were identified as belonging to DRT.

Some relevant examples were Navya's trials at the confluence district in Lyon (France), the last-mile commute at the Swiss city of Sion, and in Las Vegas offering a looped microtransit commute downtown. Also worth mentioning are EasyMile's

Typologies of uses for ASCTs		**Number of experiments**	**%**
1	First- and last-mile RLT in closed traffic	23	26
2	First- and last-mile RLT in mixed traffic	17	19
3	Microtransit RLT in closed traffic	18	20
4	Microtransit RLT in mixed traffic	21	24
5	First- and last-mile DRT in closed traffic	1	1
6	First- and last-mile DRT in mixed traffic	1	1
7	Microtransit DRT in closed traffic	4	4.5
8	Microtransit DRT in mixed traffic	4	4.5
Total		**89**	**100%**

Source: [20].

Table 1.
Classification of experimentations with typologies of use.

efforts in Calgary and Edmonton (Canada), offering a looped closed-road microtransit commute for tourists, as well as their deployments in Germany (Ioki project), in Norway (Kolumbus project), and in the United States (GoMentum station project).

As for DRT examples, it is worth highlighting the shuttle service offered by Navya on Paris Charles de Gaulle Airport, connecting passengers from the RER train station to the airport's terminals and the Schöneberg's district experimentation in Berlin, where Local Motors (American Shuttle Manufacturer) has tested an on-demand ride-hailing service based on their shuttle's (OLLI) artificial intelligence system.

As a final analysis on the benchmark study, a framework was drawn with the aim of identifying the main common stakeholders for all experimentations by listing four main groups, (1) private entities, (2) public entities, (3) communities, and (4) others, and also by describing how different forms of value (financial, usage, research, and data) is likely to flow among to and from stakeholders.

By doing so, the author concluded that ASCTs are embedded in a business ecosystem [22, 23] in which growth and evolution depend on the synergy and value flows among all stakeholders. Therefore, under the upright policy, large urban centers could strongly benefit from the introduction of ASCTs [24]. For the authors, besides of being the first- and last-mile connection to mass transit, ASCTs could compete with automobiles by price and be more effective than traditional public transport busses (by taking 10 instead of 150 passengers), being on-demand instead of on-schedule and moving on flexible routes instead of fixed ones.

Nevertheless several issues and assumptions regarding the social and economic impacts of these experimentations could be raised. In this sense, over the next section of this chapter, an in-depth analysis of three selected European project is made in order to address the aforementioned impacts with the aim of learning from these existing experimentations and understanding what could be applied and/or avoided in future projects and deployments.

3. Social and economic analysis of ASCT

For now, the technology itself is no longer the major hindrance for autonomous shuttle implementation, and the deployment speed is very much liked to the socioeconomic impacts of ASCTs on the transport system. It is therefore essential to explore the degree to which these shuttles may prove beneficial or adverse in achieving common societal and economic goals and achieve a better understanding of how transport users and society at large perceive and value their future use.

Therefore, we analyzed more deeply three main European projects (CityMobil2, SOHJOA, and GATEway) by identifying the most relevant social and economic findings as well as understanding how such results may contribute to future projects and trials.

We selected these three projects based on four criteria, in a sense that experimentations should have been (1) finished/completed, (2) deployed in mixed traffic conditions (not on exclusively dedicated roads), (3) mid- to long-term trials and/or regular services (short-time showcases were not considered due to their limited data-gathering potential), and (4) easy accessible (with public available final reports, academic papers, news, videos, blogs, and etc.).

The CityMobil2 project was a European Commission-funded project carried out from September 2012 to August 2016. It aimed at fostering the implementation of ASCTs in seven European cities spread across six countries; the general objectives included the study of long-term socioeconomic impacts of automated mobility and the definition of a legal framework that would allow ASCTs on urban roads.

CityMobil2 covered more than 25,000 km driven by the autonomous shuttles and carried over 60,000 passengers. The project is—to date—the most extensive trial with ASCTs worldwide [24].

The physical and virtual innovation platform of autonomous last-mile urban transportation (SOHJOA project) was a Finnish consortium—funded by the European Regional Development Fund—lead by the Metropolia University of Applied Sciences, from June 2016 to May 2018. The project was the first of its kind to introduce ASCTs in mixed traffic conditions on Finnish roads. It aimed at utilizing an enterprise- and area-based approach to creating new innovations and understanding related to the use of autonomous minibusses in last-mile transportation for the benefit of both the public sector and companies. Trials were carried out on three demonstrator cities (Espoo, Helsinki, and Tamper) carrying passengers over small mixed traffic looped routes (no longer than 1 km) and having to face Finland's harsh weather conditions.

The Greenwich Automated Transport Environment (GATEway project) was a British-founded consortium—jointly funded by the government and industry—that took place from July 2014 to the end of 2017 in London at the Greenwich Peninsula. It was divided into three main experimentations: (1) an autonomous shuttle service, (2) simulations and trials on how automation is supporting accessibility for disabled people, and (3) simulations and trials with autonomous delivery of goods [25]. The shuttles' project ran during April 2017 in a 1.6 km mixed traffic route (among pedestrians and cyclists) on the Greenwich Peninsula. A total of 2310 trips carrying a total of 320 were completed yielding in a total distance covered of 3700 km. It is worth noting that different from the other two projects described here, the shuttles used on GATEway, had a capacity of transporting a maximum of four passengers at a time, while the shuttles on CityMobil2 and SOHJOA could carry two or three times more passengers.

All shuttles deployed in the three analyzed projects were in accordance with SAE level-4 vehicles [26], which, for legal and safety reasons, were only allowed to drive at slow speeds (no more than 15 km/h) and were required to have a human operator on board at all times—in case automation failed.

3.1 Social impacts of autonomous shuttles' deployments

Over the past years, there has been a large number of studies aimed to investigate the acceptance of automated vehicles by gathering social-demographic, mobility, psychological, functional-utilitarian, and symbolic-affective characteristics [24]. However, most studies have focused on AVs with steer and pedals, as a result, knowledge on the factors that drive the acceptance of ASCTs in real environmental conditions is still limited [27].

Thereby, the three sampled projects were mainly focused on gathering such social insights from users, regarding acceptance, trust, willingness to use, shuttles' interactions with pedestrians, cyclists and other vehicles, and so on. Furthermore, all three projects were consistently safety-focused. Since in order to be socially accepted (and thereby used), ASCTs will need to be reliable and safe.

The main social results obtained in the sampled trials have shown positive attitudes toward the implementation of ASCTs, with the general public perceiving the vehicles as convenient, accessible, and safe. About two-thirds of the respondents stated that they would choose an automated bus if both automated and conventional busses were available in a route [28]. Most participants were positive about the safety benefits, believing that ASCTs would be either safer than or as safe as human-driven vehicles, hence considering useful to implement ASCTs services on a permanent basis.

People's first impressions were a mix of curiosity, amusement, and an urge to try something new. As reported by [29], most users made their trips to test the shuttles, that is, occasional trips were much more frequent than systematic ones. Thereby, the provided services will have to meet people's actual mobility needs in order for ASCTs to be accepted and used.

Levels of distrust and fear have also been reported. According to GATEway's final report, building the public's confidence in the technology will be a critical factor in ASCTs' successful adoption. However, as depicted by [30], trust formation is a dynamic process that starts long before a user's first contact with the system and continues long thereafter. One good experience already enhances the personal feeling of safety considerably [31], but according to the authors, people still have hesitations about the general safety of ASCTs, in a sense that opposite to a human driver, mistakes will not likely be so easily accepted.

In this sense, quality of service was an indicator that clearly needs improvements: lower operational speeds, abrupt braking, and occasional localization problems, longer waiting times, better locations and routes, comfort complaints (limited seats), information availability, etc. are all issues that require urgent attention. In addition, providing the correct infrastructure and increasing public engagement and awareness of the vehicle's capabilities are also likely to increase the acceptance of these AVs.

Furthermore, as people become more familiar with ASCTs, the excitement and enjoyment of using them may decrease [32]. Hence, in order to maintain higher satisfaction levels, manufacturers and service providers will need to ensure that these systems perform to an optimum level and are reliable, along with optimizing their connectivity with other transport services [33].

The pathway to adoption and social acceptance of ASCTs should be incremental and iterative, providing users with hands-on experience of the systems at every stage, thereby removing unrealistic and idealized expectations which can ultimately hamper acceptance.

The private semiautomated vehicle available today and in the near future provides a level of convenience and comfort which is perhaps superior to ASCTs. Therefore, for those kinds of vehicles to be considered a serious alternative to privately owned vehicles, city authorities will need to work with manufacturers and suppliers to enable the development of some—if not all—of the current features offered by privately owned vehicles.

3.2 Economic impacts of autonomous shuttles' deployments

Results on economic impacts were not widely explored in the projects in the same way as social acceptance impacts were. Hence, robust results on economic aspects were not addressed or were not disclosed on the projects' publications. The few results found were mainly concerned with users' willingness to pay for the services and the potential to reduce fares (due to the lack of a human driver).

The majority of respondents on the CityMobil's trials were positive about ASCTs if the service was offered at a lower price (with fares compatible to those of public transport) [34].

On the GATEway project, participants stated that they would be willing to pay an average price of around £2 to use ASCTs (less than the average price of a single-journey ticket in London), also believing that the shuttles have the potential to be more economical (regarding fuel/energy consumption) and therefore better for the environment.

Results from SOHJOA also state the elimination of drivers' wages as a positive outcome. However interestingly, users were not hoping for a consequent reduction on fare costs; it was hoped that the funds saved would be used to improve the quality of the mobility services (e.g., more frequent lines, on-demand services, and operations around the clock). Thereby, the authors stated that reduced costs are not necessarily considered to improve customers' quality of life but that better services are more likely to do that.

Taking into account all the aforementioned results, an interesting discussion arises. Most of the autonomous shuttle pilots aimed mostly on (1) integrating their services into the public transportation system, (2) testing the technology and circumstances, and (3) looking at people's standpoint on autonomous vehicles—since one of the biggest concerns, along with the technical performance, is to see how the general public accepts autonomous shuttles.

Exposing commuters to the technology at very early stages on a small-scale and under controlled conditions makes it possible to gradually and slowly expose and familiarize them with ASCTs' technology [34]. In this way, the introduction of automated vehicles can be linked to the creation of realistic expectations, which have been defined as a key driver of acceptance [35].

We agree with the assertions made by the authors. However, we believe that the relevance of the economic impacts are not receiving the deserved attention and thereby are somewhat being neglected. We believe that it is indeed pivotal to understand social acceptance for proper implementation of ASCTs; however, we also advocate that economic aspects also comprises the concept of acceptance.

Nevertheless, we argue that economic aspects shall not be treated as "ceteris paribus" in the data collection of experimentations with ASCTs. Their implementation requires synergy and alignment of value flows among multiple stakeholders (e.g., shuttle manufacturers, transport operators, client cities or firms, end users, digital service providers, local transport bodies, and R&D centers). Thereby, understanding economic aspects, such as costs structures, revenue flows, taxes, subsidies, investments, etc., in this business ecosystem is fundamental for successfully implementing ASCTs.

The technological evolution of the area is evident. However, it is still necessary to understand broader aspects of the industry, such as the market factors surrounding them and other economic and managerial issues [36].

Within this context, we can identify a possible gap between such technological advancements on vehicular automation and its eventual market insertion and consolidation as business models play an extremely important role in the events that precede AVs' market introduction [37].

The ultimate success of ASCTs will only be achieved following an effective collaboration between the manufacturers and local and central government, to provide citizens with the most suitable options for each specific environment.

4. A new paradigm for collective transport: Shared on-demand mobility

While the shift toward shared AV operation could indeed have a significant impact on safety, as more than 90% of road accidents are caused by human error, a shift toward shared mobility has the potential of significantly reducing the urban space allocated to parking, as well as pollution and congestion. Combining AV and collective transport may be the solution for a more citizen friendly urban space.

Besides, commuters and city's inhabitants are more and more expecting a new type of mobility that is more sustainable but also more flexible than the everyday mobility they have been used to for a long time. If they claim to be ready to share their mobility, they also require a customized mobility. In the framework of Mobility-as-a-Service, the city's authorities have to supply their voters with shared on-demand mobility.

An evolution of this type goes beyond seeing an autonomous shuttle simply as a new product but rather as a new system, the product-service system (PSS), combining technological innovation with service and market innovation. The PSS may be developed within new business models [38].

New business models are emerging from the transformation of the mobility ecosystem largely supported by digitalization. "Digitalization is a sociotechnical process that leverages the technical process of the encoding of analog information in a digital format (digitizing) applied to broader social and institutional contexts, transforming their sociotechnical structures, thus rendering digital technologies infrastructural" [39].

In the automotive industry, the potential transformation enforced by digital innovation is inducing business model innovation that can widen horizons and business paths that may impact companies' strategies toward greater sustainability and more customer services.

> *"MaaS predicts a paradigm with service providers offering travelers easy, flexible, reliable, well-priced, and environmentally sustainable everyday travel, mixing public transport, car-sharing, car leasing, and road use, with more efficient goods shipping and delivery possibilities" ([40], p. 248). "It opens up opportunities for greater customer service and potential reductions in public subsidy for public transport service; it has the very real opportunity to match customer needs more closely to service supply" ([41], p. 90).*

Indeed, as RethinkX's report [42] states: "by 2030, within 10 years of regulatory approval of autonomous vehicles, 95% of passenger miles traveled will be served by on-demand autonomous electric vehicles owned by fleets, not individuals, in a new business model called "transport-as-a-service."

Transportation-as-a-Service (TaaS) sits at the intersection of four macro trends: the technology that offers autonomous vehicles, the social demand for energy transition that pushes electrified vehicles, the overall connectivity, and the sharing economy that transfers property to usage. It involves a shift away from ownership of modes of transport toward mobility solutions that are consumed as a service.

As Mulley ([40], p. 249) states: "one of the biggest hurdles for transition to a Mobility as a Service business model is the need for a cultural shift, away from personal car ownership and reliance, toward the multiple, often shared and public mobility offerings."

That is the reason why the transition can only be started with public transport targeting specific customers, such as younger or older people, people with reduced mobility, and disabled people. For these targets, MaaS will widely increase their mobility because it will offer "mobility-on-demand," ensuring a completely new perspective in terms of mobility for people in need, reducing costs and time spent in transportation systems, and offering comfort to those who do not want to or cannot drive.

ASCT will potentially enable communities of citizens to gain greater access to opportunities for employment, education, health, and social interaction. These social benefits are likely to be particularly relevant for travelers with additional needs, such as those who are elderly, persons with disabilities, and those living in areas that are underserved by the existing transport provision.

In fact, the most remarkable asset for MaaS will undoubtedly be the portfolio of demand-responsive services, "using big data to innovate entirely new operations models to deliver new products and services based on a closer understanding of customers' on-going needs" ([43], p. 4).

Thus, we can imagine usage scenarios for an autonomous shuttle fleet operating in an urban area, bringing both more flexibility to the users and a larger variety of

Time slot	Mobility services
6–9 AM Point-to-point	Transportation with predetermined stops for regular, fixed time mobility (employees and schoolchildren)
9–5 PM On-demand	Transportation of goods (last mile) in city centers for retailers and individuals, with booking and connection to track the delivery process in real time
	Transportation for targeted needs (people with reduced mobility, leisure centers, care centers, specific goods, etc.)
	Transportation for disabled people at set times
	Transportation for city tours and outings
5–8 PM Point-to-point	Transportation with predetermined stops for regular, fixed time mobility (employees and schoolchildren)
8–6 AM On-demand	Night transportation for specific and emergency requests (like injured or sick people, delivery, deliveries for hospitals, tourist trips, etc.) Specific requests should be privately funded (individuals, travel agencies, retailer associations, etc.)

Table 2.
Scenario for a typical week day for the autonomous fleet.

business models for the transport operator. For each use, the operator may find a specific funding: subsidiaries for citizen mobility, companies for freight, Ministry of Health for disabled people, Ministry of Education for pupils, etc.

The list of user requirements for traditional forms of transport in urban centers is quite infinite: school journeys, home-to-work journeys, transport for disabled people, last-mile delivery, transport for medical care, sports and cultural activities, tours, goods deliveries, etc.

In a prospective vision featuring a totally autonomous shuttle (level 5), mobility services operate nonstop (24/7) without additional wage costs. All users are connected to the shuttle's navigation system, which must be constantly updated to ensure the service quality.

Three scenarios are currently possible: shuttle fleet management alternating predetermined journeys with request journeys, shuttle fleet management adapted to specific weekend needs, and shuttle fleet management based entirely on request journeys.

The scenarios allow a wide range of journey requests that have to be optimized by alternating point-to-point journeys, characterized by predetermined and fixed stops supporting high passenger flows, with on-demand journeys, characterized by mobile stops supporting shifting times and specific requests (**Table 2**).

Mobility-on-demand (for people or for goods) is the cornerstone of autonomous fleet deployment that could easily be combined with conventional, regular transport management featuring predetermined stops and times. A different schedule could be available during holiday periods, with more frequent shuttles during the day for tourists or citizens' leisure activities in the city.

The optimum scenario for urban mobility involves designing shared and optimized mobility-on-demand based on a constant connection between the shuttle's navigation system and the users' digital interface (smartphone or computer) in a sharing perspective.

Shared mobility-on-demand will be designed with data mining technologies using personal data to create digital people-grouping pick-up stops and common traveling time slices that will optimize autonomous fleet management. Data management and machine learning should help integrate unanticipated requests into a global optimized system. This definitively anchors mobility within a MaaS perspective.

5. Conclusion

By aiming at forecasting the future for autonomous collective transport, we conducted a benchmark on experimentations with ASCTs worldwide; the present study identified 92 deployments, spread across 78 cities in 32 countries, being enabled by 20 different shuttles' manufacturers. Figures show that Europe is at the forefront not only in experimentation numbers but also on shuttles' R&D.

By analyzing the prevailing business models, we observed that the vast majority of experimentations fit B2C models, that is, toward public transportation with ordinary commuters consisting as the main revenue source for the service provider.

Regular-line transport system comprised the vast majority of operation models; however, demand-responsive transport and mixed models comprising both regular and on-demand models are increasing, as more countries and cities begin to allow testing and circulation of AVs. Business typologies picture a range of models including first- and last-mile mobility or microtransit commute mostly within the regular-line transport system. All experimentations are part of the Mobility-as-a-Service paradigm.

The strength of the MaaS perspective is definitely its user-centric approach that allows integrating the latest technologies and the user needs for a new urban mobility. It offers a fully personalization of services based on user profiling, vehicle's localization, and dynamic itinerary optimization. Experts agree that robomobility will provide autonomous vehicle public transport services à la carte, anytime, anywhere, and for anybody, with optimal service costs and with full integration and coordination with existing public transport services. Besides, robomobility may positively impact social inclusion and well-being and give users the opportunities for social interaction offered by shared mobility.

The new transport offer will bring various innovative in- and out-of-vehicle services for passengers that will be designed with users in mind to ensure acceptance and promote the smooth integration of these novel technologies into an everyday lifestyle. In cooperation with vehicle manufacturers, public transport operators will focus on including both users in good health and users with special needs. They will develop services and solutions to accommodate these specific needs, such as children and adults who have special requirements, people with reduced mobility, and young children and elderly people who require supervision.

Author details

Sylvie Mira Bonnardel[1*], Fabio Antonialli[2] and Danielle Attias[2]

1 Ecole Centrale Lyon, Lyon University, France

2 CentraleSupelec, Paris-Saclay University, France

*Address all correspondence to: sylvie.mira-bonnardel@ec-lyon.fr

References

[1] United Nations. 68% of the world population projected to live in urban areas by 2050. (United Nations press release). 2018. Available from: https://esa.un.org/unpd/wup/Publications/Files/WUP2018-PressRelease.pdf [Accessed: 20 July 2018]

[2] Antonialli F. International benchmark on experimentations with autonomous shuttles for collective transport. In: Proceedings of International Colloquium of Gerpisa: Paradigm Shift? The Automotive Industry in Transition; 12-14 June; Paris; 2019

[3] Melis A, Mirri S, Prandi C, Prandini M, Salomoni P, Callegati F. Crowd Sensing for smart mobility through a service-oriented architecture. In: Proceedings of the Smart Cities Conference (ISC2), 2016 IEEE International; Trento, Italy; 2016

[4] Clausen C. Niche Management of autonomous vehicles for positive environmental outcomes in Copenhagen: Evaluating the feasibility of purposive measures through scenario analysis [Master's thesis]. Lund, Sweden; Lund University; 2017

[5] Alazzawi S, Hummel M, Kordt P, Sickenberger T, Wieseotte C, Wohak O. Simulating the impact of shared, autonomous vehicles on urban mobility—A case study of Milan. EPiC Series in Engineering. 2018;**2**:94-110

[6] Attias D. The automobile world in a state of change: from the automobile to the concept of auto-mobility. In: Attias D, editor. The Automobile Revolution: Towards a New Electro-Mobility Paradigm. 1st ed. Gewerbestrasse (Switzerland): Springer International Publishing; 2017

[7] PricewaterhouseCoopers. Five Trends Transforming the Automotive Industry. PWC Report. 2018. Available from: https://eu-smartcities.eu/sites/default/files/2018-03/pwc-five-trends-transforming-the-automotive-industry.compressed.pdf [Accessed: 20 July 2018]

[8] Alazzawi S, Hummel M, Kordt P, Sickenberger T, Wieseotte C, Wohak O. Simulating the impact of shared, autonomous vehicles on urban mobility—A case study of Milan. EPiC Series in Engineering. 2018;**2**:94-110

[9] Lang N, Rübmann M, Mei-Pochtler A, Dauner T, Komiya S, Mosquet X, et al. Self-Driving Vehicles, Robo-Taxis, and the Urban Mobility Revolution. Boston Consulting Group Report. 2016. Available from: https://www.bcg.com/pt-br/publications/2016/automotive-public-sector-self-driving-vehicles-robo-taxis-urban-mobility-revolution.aspx [Accessed: 20 July 2018]

[10] Mutz F, Veronese LP, Oliveira-Santos T, de Aguiar E, Cheein FAA, De Souza AF. Large-scale mapping in complex field scenarios using an autonomous car. Expert Systems with Applications. 2016;**46**:439-462

[11] Fagnant DJ, Kockelman K. Preparing a nation for autonomous vehicles: Opportunities, barriers and policy recommendations. Transportation Research Part A: Policy and Practice. 2015;**77**:167-181

[12] Schellekens M. Self-driving cars and the chilling effect of liability law. Computer Law and Security Review. 2015;**31**(4):506-517

[13] Mira-Bonnardel S, Attias D. The autonomous vehicle for urban collective transport: Disrupting business models embedded in the smart city revolution. In: Proceedings of International Colloquium of Gerpisa: Who Drives the

Change? New and Traditional Players in the Global Automotive Sector; São Paulo; 2018

[14] Mira-Bonnardel S, Attias D. Autonomous vehicle for flexible public transportation systems: Towards a shared on demand mobility. In: Proceedings of International Colloquium of Gerpisa: Paradigm shift? The Automotive Industry in Transition; 12-14 June; Paris; 2019

[15] Harris R. Shuttle Diplomacy: The Path to Deployment. Transportation as a Service Magazine. 2018. Available from: https://www.publishing.ninja/V4/page/4572/245/270/1 [Accessed: 30 September 2018]

[16] Antonialli F. International benchmark on experimentations with autonomous shuttles for collective transport. In: Proceedings of International Colloquium of Gerpisa: Paradigm Shift? The Automotive Industry in Transition; 12-14 June; Paris; 2019

[17] Harris R. Shuttle Diplomacy: The Path to Deployment. Transportation as a Service Magazine. 2018. Available from: https://www.publishing.ninja/V4/page/4572/245/270/1 [Accessed: 30 September 2018]

[18] AVENUE. Objectives of the Avenue Project. 2018. Available from: https://h2020-avenue.eu/content/objectives-avenue-project [Accessed: 22 June 2019]

[19] Mira-Bonnardel S, Attias D. The autonomous vehicle for urban collective transport: Disrupting business models embedded in the smart city revolution. In: Proceedings of International Colloquium of Gerpisa: Who Drives the Change? New and Traditional Players in the Global Automotive Sector; São Paulo; 2018

[20] Antonialli F. International benchmark on experimentations with autonomous shuttles for collective transport. In: Proceedings of International Colloquium of Gerpisa: Paradigm Shift? The Automotive Industry in Transition; Paris, 12-14 June; 2019

[21] Fluhr D. Navya vs EasyMile—Who will Conquer the Autonomous E-Shuttle Market? Smart Mobility Hub. 2017. Available from: http://smart-mobility-hub.com/navya-vs-easymile-who-will-conquer-the-autonomous-e-shuttle-market/ [Accessed: 30 September 2018]

[22] Moore JF. Predators and prey: A new ecology of competition. Harvard Business Review. 1993;**71**(3):75-86

[23] Moore JF. The rise of a new corporate form. The Washington Quarterly. 1998;**21**(1):167-181

[24] Ainsalu J et al. State of the art of automated buses. Sustainability. 2018;**10**(3118):2-34

[25] Cuerden R. GATEway: This is Just the Beginning—Positioning the UK at the Forefront of Automated Mobility. GATEway Project Summary Report. 2018. Available from: https://gateway-project.org.uk/wp-content/uploads/2018/06/D1.3_GATEway-Project-Final-Report-brochure.pdf [Accessed: 29 March 2019]

[26] SAE. Taxonomy and Definitions for Terms Related to Driving Automation Systems for On-Road Motor Vehicles J3016_201609, Society of Automotive Engineers. 2016. Available from: https://www.sae.org/standards/content/j3016_201609/ [Accessed: 05 July 2018]

[27] Nordhoff S et al. User acceptance of driverless shuttles running in an open and mixed traffic environment. In: Proceedings of the ITS European Congress; Strasbourg, France. 2017

[28] Alessandrini A. Final Report Summary—CityMobil2 (Cities

Demonstrating Cybernetic Mobility). European Commission—CORDIS. 2016. Available from: https://cordis.europa.eu/project/rcn/105617/reporting/en [Accessed: 03 April 2019]

[29] Merat N, Madigan R, Nordhoff S. Human Factors, User Requirements, and User Acceptance of Ride-Sharing in Automated Vehicles. International Transport Forum—Discussion Paper. 2017. Available from: https://www.google.com.br/url?sa=t&rct=j&q=&esrc=s&source=web&cd=2&ved=2ahUKEwi97onnuPXhAhWjAmMBHV0vANoQFjABegQIBBAC&url=https%3A%2F%2Fwww.itf-oecd.org%2Ffile%2F16355%2Fdownload%3Ftoken%3DwzML3Ngi&usg=AOvVaw3u20zlsLVA-aIGHQmK0810 [Accessed: 18 April 2019]

[30] Ekman F, Johansson M, Sochor J. Creating appropriate trust in automated vehicle systems: A framework for HMI design. IEEE Transactions on Human-Machine Systems. 2017;**48**(1):95-101

[31] Salonen AO, Haavisto N. Towards autonomous transportation. Passengers' experiences, perceptions and feelings in a driverless shuttle bus in Finland. Sustainability. 2019;**11**(588):1-19

[32] Sessa C, Alessandrini A, Flament M, Hoadley S, Pietroni F, Stam D. The socio-economic impact of urban road automation scenarios: CityMobil2 participatory appraisal exercise. In: Meyer G, Beiker S, editors. Road Vehicle Automation 3. Gewerbestrasse, Switzerland: Springer International Publishing; 2016

[33] Piao J, McDonald M, Hounsell N, Graindorge M, Graindorge T, Malhene N. Public views towards implementation of automated vehicles in urban areas. Transportation Research Procedia. 2016;**14**:2168-2177

[34] Nordhoff S et al. User acceptance of driverless shuttles running in an open and mixed traffic environment. In: Proceedings of the ITS European Congress; Strasbourg, France. 2017

[35] Nees MA. Acceptance of self-driving cars: An examination of idealized versus realistic portrayals with a self- driving car acceptance scale. Proceedings of the Human Factors and Ergonomics Society Annual Meeting. 2016;**60**(1):1449-1453

[36] Gandia RM, Antonialli F, Cavazza BH, Neto AM, Lima DAD, Sugano JY, et al. Autonomous vehicles: Scientometric and bibliometric review. Transport Reviews. 2018;**38**:1-20

[37] Cavazza BH et al. Management and business of autonomous vehicles: A systematic integrative bibliographic review. International Journal of Automotive Technology and Management. 2019;**19**(1-2):31-54

[38] Jilttrapirom P, Caiati V, Feneri A, Ebrahimigharehbaghi S, Alonso-Gonzalez MJ, Narayan J. Mobility as a service: A critical review of definitions, assessments of schemes, and key challenges. Urban Planning, Lisbon. 2017;**2**(2):13-25

[39] Kaiser C, Stocker A, Viscusi G. Digital vehicle ecosystems and new business models: An overview of digitalization perspectives. i-KNOW 2017; 11-12 October; Graz, Austria; 2017

[40] Mulley C. Mobility as a services—Does it have critical mass? Transport Reviews, Oxfordshire, 37. 2017;(3):247-251

[41] Hensher DA. Future bus transport contracts under a mobility as a service (MaaS) regime in the digital age: Are they likely to change? Transportation Research Part A: Police and Practice, Amsterdam. 2017;**98**:86-96

[42] Rethink. Disruption, Implications and Choices Rethinking Transportation 2020-2030. The Disruption of

Transportation and the Collapse of the Internal-Combustion Vehicle and Oil Industries. 2017. 77p. Available from: https://static1.squarespace.com/static/585c3439be65942f022bbf9b/t/591a2e4be6f2e1c13df930c5/1494888038959/RethinkX+Report_051517.pdf

[43] Graham G, Meriton R, Tew B, Hennelly P. How is big data transforming operations models in the automotive industry: A preliminary investigation. In: International Conference on Emerging Networks and Systems Intelligence. 8. Venice: University of Leeds; 2016. pp. 1-5

Section 2

System Controls

Chapter 3

Stator Winding Fault Diagnosis of Permanent Magnet Synchronous Motor-Based DTC-SVM Dedicated to Electric Vehicle Applications

Fatma Ben Salem

Abstract

Electric vehicles (EVs) have several advantages such as energy efficiency, virtually lack of pollution, and the availability of electric energy through electric distribution systems. Among the major components of an EV is the block control of the AC motor. Many modern high-performance control technologies are being used in EV controllers, such as direct torque control-based space vector modulation (DTC-SVM). Given that permanent magnet synchronous motor (PMSM) has a small volume, a lightweight, and a high reliability, attracting the interest of EV industrial designers. The fault diagnosis about the PMSM problems is essential because the PMSM has a huge influence on the safety and economic efficiency with the vehicles. In this background, the chapter concerns the modeling and simulation of an electric vehicle drive system including a PMSM-based DTC-SVM without and with stator winding fault. The performance of the motor under both the normal and faulty conditions is simulated. For the diagnosis of the fault, the chapter proposes the technique based on the motor current spectrum analysis.

Keywords: permanent magnet synchronous motor, electric vehicles, DTC-SVM, stator winding fault, diagnosis, fault signature extraction

1. Introduction

Electric vehicles (EVs) have several advantages such as energy efficiency, virtually lack of pollution, and the availability of electric energy through electric distribution systems [1–4]. Among the major components of an electric vehicle is the controlling block of the AC motor. The controller should be designed to make the drive system robust and efficient on both dynamic and steady-state performances.

Many modern high-performance control technologies are being used in EV controllers, such as direct torque control-based space vector modulation (DTC-SVM) [5–12]. The advantage of the DTC-SVM methodology is that it has all the merits of DTC strategy [13–16] by adding the fact of having a fixed switching frequency, which guarantees low vibrations in order to provide passenger comfort.

Permanent magnet synchronous motor (PMSM) becomes more suitable in wide speed range application due to their high efficiency and wide constant power speed range. In fact, the absence of rotor windings and therefore the absence of external excitation increase the efficiency of the PMSM and greatly reduce its maintenance cost. In addition, the drop in the price of permanent magnets promotes a strong use of this type of machine particularly in the field of electric vehicle applications [17–20]. In fact, such motor has a small volume, lightweight, and high reliability attracting the interest of EV industrial designers.

So the fault diagnosis about the PMSM problems is essential because the PMSM has a huge influence on the safety and economic efficiency of the vehicles. The stator winding fault is a common fault that has interested researchers for many years [21–27]. It is resulting from the degradation of the inter-turn interphase and main isolation of the motor winding. This internal fault will increase the torque ripples that deteriorate machine performance. However such fault can be rapidly propagated to motor stator turn since it makes a large circulating current in the shorted path yielding excessive heat. The adverse condition of heat will occasionally lead to the progressive deterioration and eventual breakdown of winding insulation in the phase of the motor.

Several methods have been used to detect and estimate the type and the degree of this fault in PMSM. The motor signature current analysis is one of the most common online methods for fault detection. In fact, it is widely used in the field of PMSM fault diagnosis, because it is fast and does not need any specific model. This method uses spectral analysis techniques by comparing the healthy to the faulty case; faults can be detected.

In this chapter, as for signal processing tools, fast Fourier transform (FFT) is used to extract the frequency characteristics of the stator current signal. This method will be proposed to detect whether a PMSM, under DTC-SVM traction system, is healthy or faulty.

The chapter is organized as follows. Electric vehicle propulsion system scheme is developed in Section 2. In Section 3, the mathematical model of PMSM in (a, b, c) frame in healthy conditions is introduced and which is followed by Section 4, the PMSM with inter-turn fault modeling. Then, the design of a conventional DTC-SVM traction system is presented. Simulation result, considering steady-state operation, is discussed in Section 6.

2. Electric vehicle propulsion system scheme

The major electric propulsion system of EV consists of motor, controller, power source, charger, transmission device, and wheels (**Figure 1**) [28]. The proposed motor drive includes the electric motor (PMSM), power converter (three-phase voltage inverter), energy storage (battery), and electronic controller (DTC-SVM). These components are the core of the EV propulsion system.

The tractive effort (F_{te}) is the force propelling the vehicle forward, transmitted to the ground through the drive wheels. Consider a vehicle of mass M_v, proceeding at a velocity V_v, as in **Figure 2**. The force propelling the vehicle forward, the tractive effort, has to overcome the following forces:

- The rolling resistance force: $F_{rrol} = \mu M_v g$

where μ is the coefficient of rolling resistance ($\mu = 0.005$ for tires developed especially for electric vehicles), M_v is the vehicle mass, and g is the gravitational constant

- The aerodynamic drag: $F_{ad} = \frac{1}{2}\rho A C_d V_v^2$

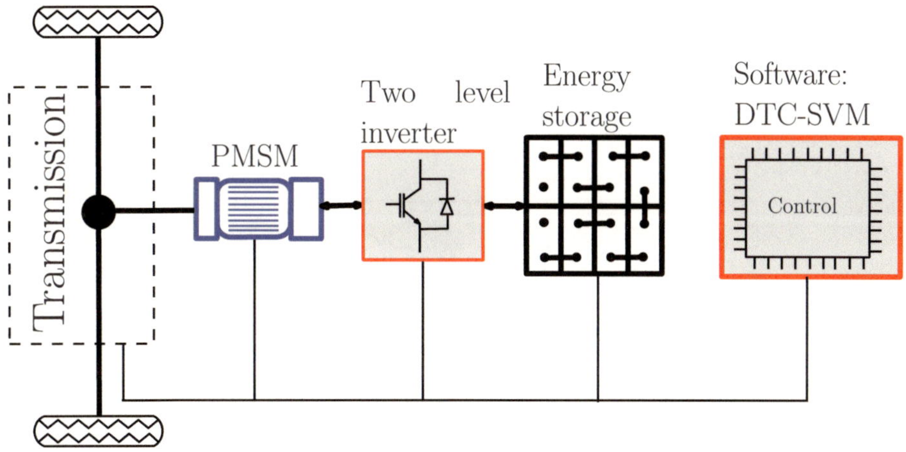

Figure 1.
Components of electric vehicle propulsion system.

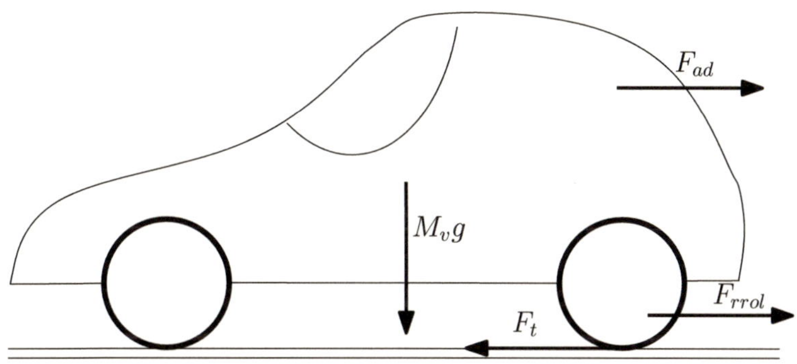

Figure 2.
The forces acting on a vehicle.

where ρ is the air density, A is the frontal area, C_d is the drag coefficient, and V_v is the vehicle speed

- The hill climbing force: $F_{hc} = M_v g sin\alpha$

where α is the angle of a slope for a vehicle climbing a hill (in the case of **Figure 2** $\alpha = 0$).

F_{te} is equal to the sum of the resistance forces [28, 29], as shown in the sequel equation:

$$F_{te} = F_{rrol} + F_{ad} + F_{hc} \tag{1}$$

2.1 Voltage source inverter

The DC voltage made constant by the rectifier is delivered by the battery to the inverter input, which thanks to controlled transistor switches, converts this voltage to three-phase AC voltage signal with wide-range variable voltage amplitude and frequency.

The inverter one leg consists of two transistor switches. A simple transistor switch consists of feedback diode connected in antiparallel with transistor.

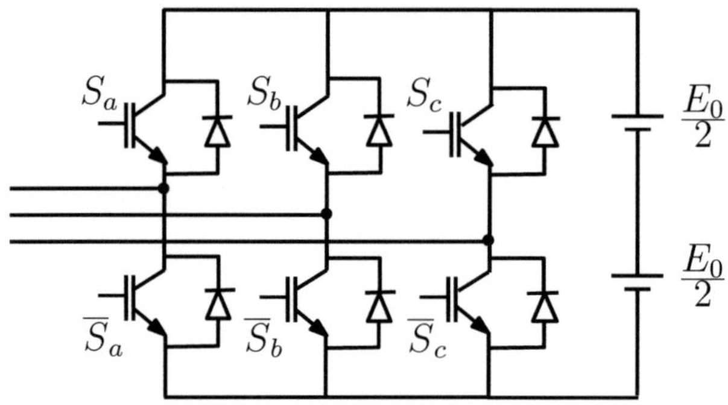

Figure 3.
Two-level three-phase voltage inverter.

Feedback diode conducts current when the load current direction is opposite to the voltage direction.

Assuming that the power devices are ideal, when they are conducting, the voltage across them is zero and they present an open circuit in their blocking mode. Therefore, each inverter leg can be represented as an ideal switch. It gives the possibility to connect each of the three motor phase coils to a positive or negative voltage of the DC link (E_0).

Considering a two-level inverter, presented by **Figure 3**, the voltage vector of the three-phase voltage inverter is represented as follows:

$$\vec{V_s} = \sqrt{\frac{2}{3}}\left[S_a + S_b e^{j\frac{2\pi}{3}} + S_c e^{j\frac{4\pi}{3}}\right] \tag{2}$$

where S_a, S_b, and S_c are three-phase inverter switching functions, which can take a logical value of either 0 or 1.

3. Mathematical model of PMSM in (a, b, c) frame

An accurate dynamic model of the PMSM is necessary to study the dynamic behavior of the machine. Indeed, the dynamic behavior of a PMSM is described in terms of space variables as follows:

$$V_{abc} = R_s I_{abc} + L_s \frac{d}{dt} I_{abc} + \frac{d}{dt}\Phi_{mabc} \tag{3}$$

considering $V_{abc} = [v_a\ v_b\ v_c]^T$, $I_{abc} = [i_a\ i_b\ i_c]^T$, and

$$\frac{d\Phi_{mabc}}{dt} = \Phi_m \omega_r \left[\cos(\theta_r)\ \cos\left(\theta_r - \frac{2\pi}{3}\right)\ \cos\left(\theta_r + \frac{2\pi}{3}\right)\right]^T$$

where V_{abc}, I_{abc}, and Φ_m denote the stator voltages, the stator currents, and the permanent magnetic flux amplitude, respectively. R_s and L_s denote the stator resistance and inductance matrix, respectively.

$$\Phi_{mabc} = \Phi_m \begin{bmatrix} \sin(\theta_r) \\ \sin\left(\theta_r - \frac{2\pi}{3}\right) \\ \sin\left(\theta_r + \frac{2\pi}{3}\right) \end{bmatrix} \tag{4}$$

since ω_r=$N_p\omega_m$, ω_r and ω_m torque pulsation and mechanical pulsation, respectively. θ_r is the angular position of the rotor with respect to the magnetic axis of the phase (a) of the stator, θ_r= $N_p\theta_m$, θ_m denotes the mechanical angular position of the rotor, and N_p is the pole pair number.

The electromagnetic torque can be expressed as follows:

$$T_{em} = \frac{e_a i_a + e_b i_b + e_c i_c}{\Omega_m} \tag{5}$$

where e_a, e_b, and e_c are the emf of the phases a, b, and c and $\Phi_{mabcf} = \Phi_m[e_a\ e_b\ e_c]$. Then referring to Eq. (4), the developed equation of electromagnetic torque becomes

$$T_{em} = \frac{\Phi_m\left(\sin(\theta_r)i_a + \sin\left(\theta_r - \frac{2\pi}{3}\right)i_b + \sin\left(\theta_r + \frac{2\pi}{3}\right)i_c\right)}{\Omega_m} \tag{6}$$

The mechanical part of the machine is described by

$$J\frac{d}{dt}\Omega_m = T_{em} - T_l \tag{7}$$

where J is the motor inertia, T_{em} is the electromagnetic torque, and T_l is the load torque.

4. Modeling of the PMSM with inter-turn fault

An inter-turn fault represents an insulation failure between two windings in the same phase of the stator [23].

The representation of a PMSM with stator winding turn fault at phase (a) is shown in **Figure 4**; a_2 and a_1 represent the shorted turns and the healthy turns, respectively. The fault is modeled by a small resistance R_f connected across the shorted turns. In fact, R_s denotes the inter-turn short circuit resistance.

To represent the impact of the defect, a new parameter σ is introduced. Parameter σ is defined as the ratio between the number of shorted turns N_{cc} and the total number of turns N_s.

The resistances of the sub-winding a_1 and the shorted sub-winding a_2 are noted by R_{a1} and R_{a2}, respectively; they are proportional to the number of turns of the involved parties. Therefore, we can express them according to the resistance R_a and the coefficient σ. So we have

$$R_{a_1} = (1-\sigma)R_a, \quad R_{a_2} = \sigma R_a \tag{8}$$

By defining the electrical quantities of the new circuit with the index "f", and refering to **Figure 5** the new equations of the motor voltages are reformulated as follows:

$$\begin{aligned} V_a =&(R_{a_1} + R_{a_2})i_a + (L_{a_1} + L_{a_2} + 2M_{a_1a_2})\frac{di_a}{dt} + (M_{a_1b} + M_{a_2b})\frac{di_b}{dt} \\ &+ (M_{a_1c} + M_{a_2c})\frac{di_c}{dt} + (e_a + e_f) - R_{a_2}i_f - (L_{a_2} + M_{a_1a_2})\frac{di_f}{dt} \\ V_b =&R_s i_b + L\frac{di_b}{dt} + (M_{a_1b} + M_{a_2b})\frac{di_a}{dt} + M\frac{di_c}{dt} + e_b - M_{a_2b}\frac{di_f}{dt} \\ V_c =&R_s i_c + L\frac{di_c}{dt} + (M_{a_1c} + M_{a_2c})\frac{di_a}{dt} + M\frac{di_b}{dt} + e_c - M_{a_2c}\frac{di_f}{dt} \end{aligned} \tag{9}$$

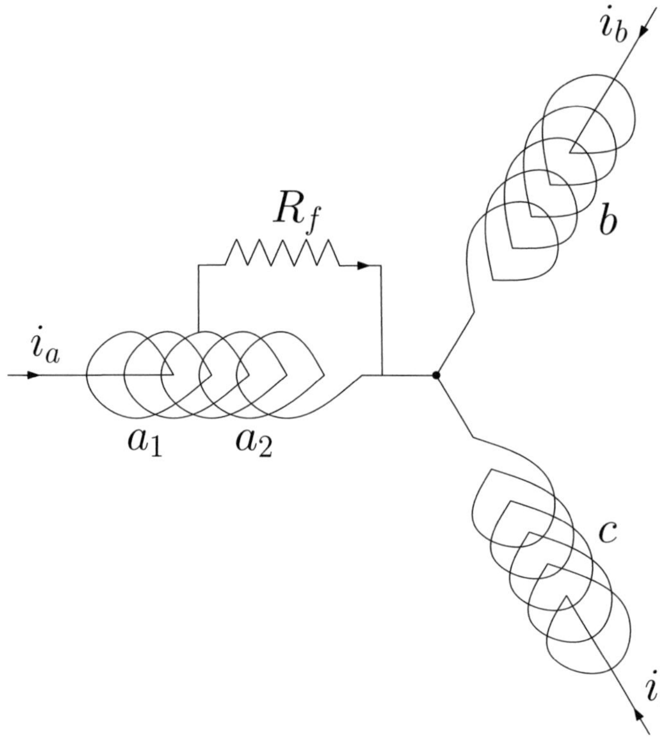

Figure 4.
Equivalent circuit of PMSM under inter-turn fault in phase (a).

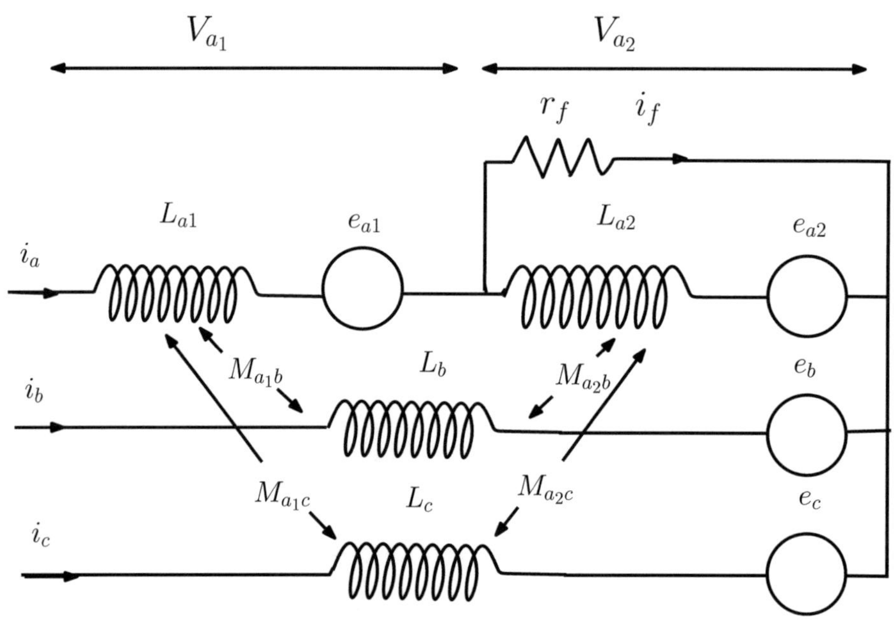

Figure 5.
Schematic representation of an insulation fault between turns on a phase of the stator.

where L and M denote the phase self-inductance and M is the mutual inductance between phase windings of healthy PMSM, respectively. And considering $L_{a_1} = (1-\sigma)^2 L_a$, $L_{a_2} = \sigma^2 L_a$ and $M_{a_1a_2} = \sigma(1-\sigma)L_a$.

5. Design of a conventional DTC-SVM traction system

The disadvantages of basic DTC is obvious: torque and flux ripples, deteriorated performance at low speed, and uncontrolled switching frequency of the inverter. For the defects of basic DTC, much works have been made over the past few decades. DTC combined with space vector modulation (DTC-SVM) for PMSM is to accomplish constant switching frequency of the inverter in addition to obtain the desired torque and stator flux with little ripples by synthesizing an appropriate voltage space vector through SVM, which is more accurate than that of basic DTC to compensate the error of desired and actual stator flux.

5.1 Flux reference coordinate computing

The relationship between the voltage, current, and stator flux vectors of a PMSM is given by

$$\frac{d}{dt}\overline{\Phi}_s = \overline{V}_s - R_s\overline{I}_s \tag{10}$$

where R_s is the stator resistance.

The reference of stator flux amplitude and phase is stated using *Concordia* quantities as

$$|\Phi_s^*| = \sqrt{\phi_{\alpha s}^{*2} + \phi_{\beta s}^{*2}}, \quad \theta_s^* = \arctan\left(\frac{\phi_{\beta s}^*}{\phi_{\alpha s}^*}\right) \tag{11}$$

5.2 Voltage reference coordinate computing

The coordinates of references of voltage vectors $v_{\alpha s}^*$ and $v_{\beta s}^*$ in (α, β) frame are determined by the following equations:

$$\begin{cases} v_{\alpha s}^* = \dfrac{\phi_{\alpha s}^* - \phi_{\alpha s}}{T_{em}} + R_s i_{\alpha s} \\ v_{\beta s}^* = \dfrac{\phi_{\beta s}^* - \phi_{\beta s}}{T_{em}} + R_s i_{\beta s} \end{cases} \tag{12}$$

These vectors are introduced to the SVM block, which use them to control the inverter switches (S_a, S_b, S_c).

5.3 SVM block design

The SVM technique refers to a special switching scheme of the six power transistors of a two-level three-phase voltage inverter. In fact, it uses eight sorts of different switch modes of the inverter to control the stator flux to advance the reference flux circle. Eight types of switch modes stand for eight space voltage vectors that contain six active voltage vectors and two zero voltage vectors as shown in **Figure 6a** [30, 31].

Voltage vectors, created by a three-phase PWM inverter, divide the space vector plane into six sectors: $(i-1)\frac{\pi}{3} < S_i \leq i\frac{\pi}{3}$, $i = 1, \ldots, 6$. The determination of the sector, where the reference voltage vector is located, is done according to

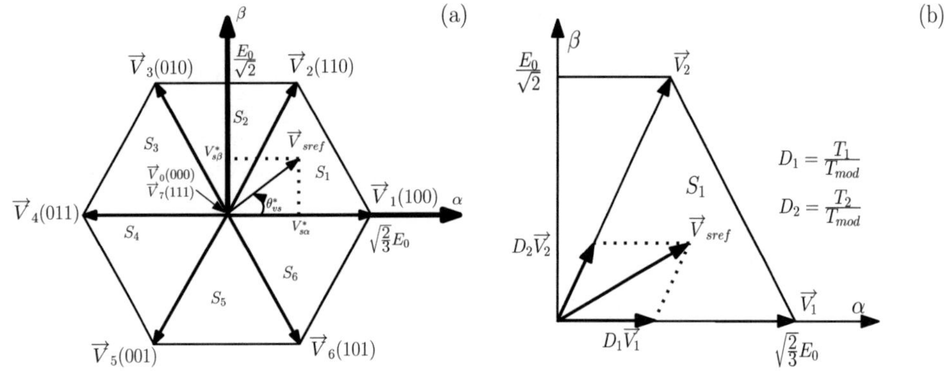

Figure 6.
(a) Basic switching vectors and sectors. (b) Projection of the reference voltage vector on two adjacent vectors.

$$\theta_s^* = \arctan\left(\frac{v_{\beta s}^*}{v_{\alpha s}^*}\right) \tag{13}$$

The SVM's principle is to project the desired stator voltage vector $\vec{V}_{sref}$ on the two adjacent voltage vectors $\vec{V}_i$ and $\vec{V}_{i+1}$ corresponding to two switching states of the inverter. Values of these projections provide the determination of desired commutation times T_i and T_{i+1} and correspond to two nonzero switching states of the inverter. To maintain the constant commutation frequency, in the case where $T_i + T_{i+1} \leq T_{mod}$, a zero state of the inverter is applied during the rest of the period T_{mod}, i.e., $T_0 = T_{mod} - (T_i + T_{i+1})$. In what follows, the study will be limited to sector S_1, as presented in **Figure 6b**. As the reference voltage vector $\vec{V}_{sref}$ is in sector S_1, it can be compounded by the active voltage vectors V_1 and V_2. The projection on these adjacent vectors gives the following expression:

$$\begin{aligned} \vec{V}sref &= V_{s\alpha}^* + jV_{s\beta}^* \\ &= \frac{T_1}{T_{mod}}\vec{V}_1 + \frac{T_2}{T_{mod}}\vec{V}_2 \end{aligned} \tag{14}$$

where $T_{mod} = T_1 + T_2 + T_0$. D_1 and D_2 are duties relative to voltages V_1 and V_2. In sector S_1, expressions of the voltage vectors are

$$\begin{cases} \vec{V}_1 = \sqrt{\frac{2}{3}}E_0(\cos 0 + j\sin 0) = \sqrt{\frac{2}{3}}E_0 \\ \vec{V}_2 = \sqrt{\frac{2}{3}}E_0\left(\cos\frac{\pi}{3} + j\sin\frac{\pi}{3}\right) \end{cases} \tag{15}$$

Expressions of T_1 and T_2 are detailed in the sequel:

$$\begin{cases} T_1 = \left(\sqrt{\frac{3}{2}}V_{s\alpha}^* - \frac{1}{\sqrt{2}}V_{s\beta}^*\right)\frac{T_{mod}}{E_0} \\ T_2 = \sqrt{2}V_{s\beta}^*\frac{T_{mod}}{E_0} \end{cases} \tag{16}$$

Consequently, the expressions of the duties are given as follows:

$$\begin{cases} D_1 = \left(\sqrt{\frac{3}{2}}V_{s\alpha}^* - \frac{1}{\sqrt{2}}V_{s\beta}^*\right)\frac{1}{E_0} \\ D_2 = \sqrt{2}V_{s\beta}^*\frac{1}{E_0} \end{cases} \quad (17)$$

The time duration of each nonzero vector is divided equally into two parts, the time duration of zero vectors is distributed equally to $\vec{V}_0$ and $\vec{V}_7$, and thus the switching sequence of space vector is $\vec{V}_0, \vec{V}_1, \vec{V}_2, \vec{V}_7, \vec{V}_7, \vec{V}_2, \vec{V}_1, \vec{V}_0$ during the modulation period.

The duties of each phase of the inverter are calculated as follows:

$$\begin{cases} D_a = D_1 + D_2 + \frac{1}{2}D_0 \\ D_b = D_2 + \frac{1}{2}D_0 \\ D_c = \frac{1}{2}D_0 \end{cases} \quad (18)$$

As $D_1 + D_2 + D_0 = 1$, duties, given by the system of Eq. (18), turn to be

$$\begin{cases} D_a = \frac{1}{2}\left(1 + \sqrt{\frac{3}{2}}\frac{V_{s\alpha}^*}{E_0} + \frac{1}{\sqrt{2}}\frac{V_{s\beta}^*}{E_0}\right) \\ D_b = \frac{1}{2}\left(1 - \sqrt{\frac{3}{2}}\frac{V_{s\alpha}^*}{E_0} + \frac{1}{\sqrt{2}}\frac{V_{s\beta}^*}{E_0}\right) \\ D_c = \frac{1}{2}\left(1 - \sqrt{\frac{3}{2}}\frac{V_{s\alpha}^*}{E_0} - \frac{1}{\sqrt{2}}\frac{V_{s\beta}^*}{E_0}\right) \end{cases} \quad (19)$$

5.4 Block diagram of the DTC-SVM

The DTC-SVM block diagram retains all the advantages of the conventional DTC, such as no coordinate transformation, robust to motor parameters, etc. Moreover, a SVM block is used to generate the pulses for the inverter in order to promise a fixed commutation frequency.

Figure 7 illustrates the block diagram of the proposed traction control system, and the overall control scheme is set as follows:

- Including two major loops: the torque control loop and the flux control loop. In fact, the flux and torque are directly controlled individually. The basic principle of the direct torque control is to destine the torque error and the flux error in hysteresis bands by properly choosing the switching states of the inverter.

- The speed command is compared with the estimating speed to compute the speed error. Then, the speed error is processed by the PI speed controller to obtain the torque command. On the other hand, the flux command is compared to the estimated flux.

- The errors ΔT_{em} and $\Delta\Phi_s$ go through the hysteresis controllers and the vector selection table to generate the required switching states and therefore the desired voltage vector.

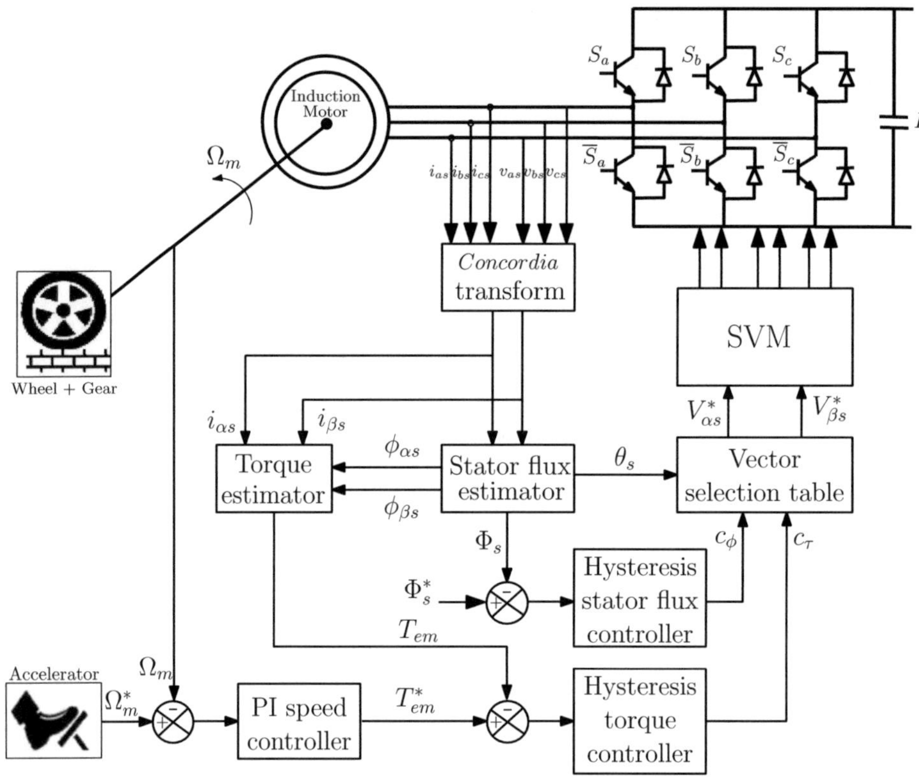

Figure 7.
Block diagram of SVM-DTC PMSM drive.

- The controllers produce the voltage command vector; an appropriate space voltage vector can be generated with SVM.

6. Simulation end discussion

Parameters of the PMSM are listed in **Table 1**. It has the following ratings: 220 V, 10 kW, and 1470 rpm at 50 Hz.

The parameters of the electric vehicle model are given in **Table 2**.

6.1 Case study

We consider a vehicle traveling along a straight horizontal road. For $F_{hc} = 0$, the tractive effort is reduced to $F_{te} = F_{rrol} + F_{ad}$, and the load torque can be expressed as $T_r = R_g r(F_{rrol} + F_{ad}) =$, where $R_g = \frac{1}{G}$ and G denotes the gear ratio.

After development the load torque is

$$T_l \;=\; K_1 + K_2 V_v^2 \tag{20}$$

$R_s = 0.29\Omega$	$L_s = L_r = 50$ mH	$N_p = 2$
$R_r = 0.38\Omega$	$M = 47.3$ mH	$J = 0.5$ Kg.m^2

Table 1.
PMSM parameters.

$r = 0.3m$	$A = 1m^2$
$M_v = 400$ Kg	$C_d = 0.19$
$G = 0.9$	$\rho = 1.2$ Kg/m^3

Table 2.
The electric vehicle model parameters.

where $K_1 = R_g r \mu M_v g$ and $K_2 = \frac{1}{2} R_g r \rho A C_d V_v^2$.
The relationship between vehicle speed and motor speed is given by

$$V_v = R_g r \ \Omega_m \tag{21}$$

Simulation tests were performed in healthy and faulty conditions, with the motor driving a load torque, referring to Eqs. (20) and (21), and can be expressed as follows:

$$T_l \quad = \quad K_1 + K_2 (R_g r)^2 \Omega_m^2 \tag{22}$$

The magnitude of the torque and flux hysteresis bands is 0.01 N.m and 0.001 Wb, respectively.

6.2 Steady-state operation analysis

Figure 8 shows several features of the PMSM drive at steady-state operation considering a speed $\Omega_m = 750$ rpm. Waveforms presented in the left side (subscript "1") of **Figure 8** represent results yielded by the PMSM under DTC-SVM without fault. The middle side (subscript "2") of **Figure 8** represents the results yielded by the PMSM under DTC-SVM considering 5% turns are shorted. The right side

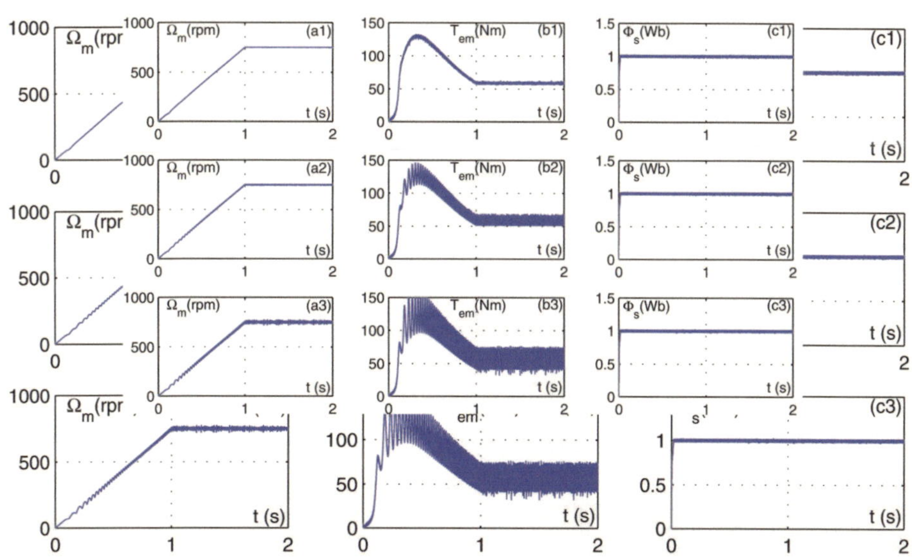

Figure 8.
(subscript "1") Considering the PMSM under DTC-SVM without fault, (subscript "2") considering the PMSM under DTC-SVM allowing for 5% shorted turns, and (subscript "3") considering the PMSM under DTC-SVM allowing for 25% shorted turns. Legend: (a) speed and its reference, (b) electromagnetic torque, and (c) stator phase current.

(subscript "3") of **Figure 8** represents the results yielded by PMSM under DTC-SVM considering 25% turns are shorted.

From the analysis of the previous steady-state simulation results, one can remark the following:

- The speed follows its reference despite the presence of defect. It is clear that in the case of machine with short circuit fault, the speed oscillates around its reference with a ripple amplitude that increases when the percentage of turns shorted increases.

- It is seen in **Figure 8** when 5 and 25% turns are shorted, respectively, in one of the phase (a) windings, the increase of torque ripples is notable.

- The stator flux shapes, in healthy and faulty conditions, are quite similar.

6.3 Signature extraction based on the analysis of the fault effect

From the analysis of **Figure 9**, one can notice that a short circuit fault between turns on phase (a) increases the amplitude of the second harmonic which was too low in the spectrum of the healthy machine. This harmonic introduced by the defect presents an amplitude that evolves the rate of defect in the sense that this harmonic sees its amplitude increases in a proportional way with the increase of the number of turns short circuited on the phase (a). When we increase the rate of the short circuit, the amplitude of this harmonic has increased in value.

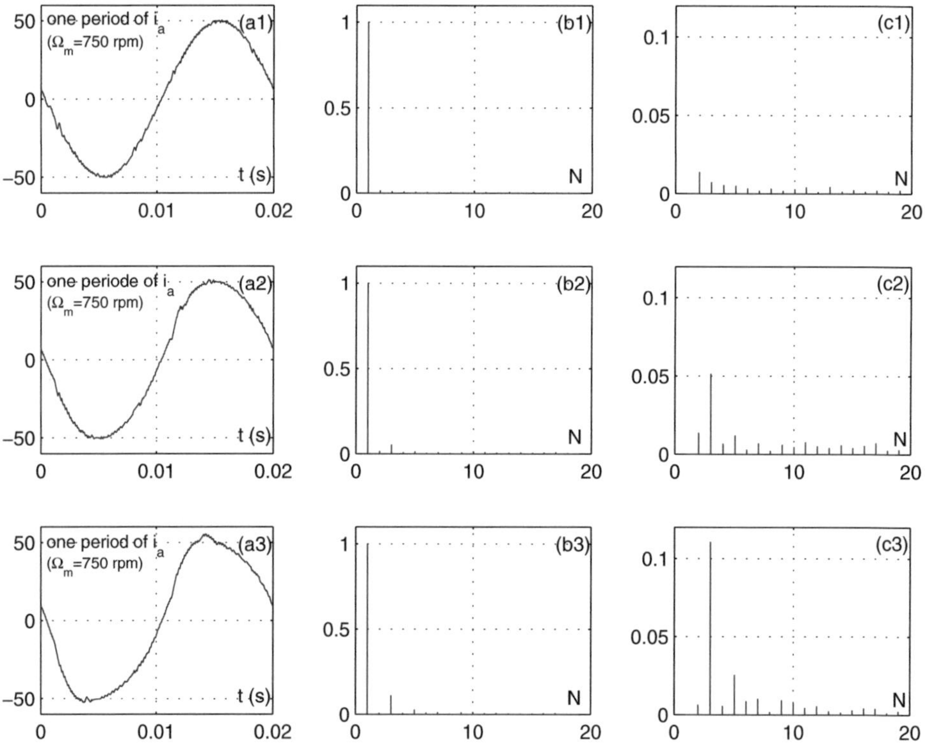

Figure 9.
(a) One period of i_{as}, (b) reduced spectrum of i_{as} with respect to fundamental, and (c) harmonic of i_{as} (fundamental = 1). (subscript "1") considering the PMSM under DTC-SVM without fault, (subscript "2") considering the PMSM under DTC-SVM allowing for 5% shorted turns, and (subscript "3") considering the PMSM under DTC-SVM allowing for 25% shorted turns.

In light of this analysis, it clearly appears that, in general, the harmonic amplitude increases in an apparent way when there is a short circuit fault.

7. Conclusion

The chapter focused on a study and diagnosis of the PMSM under DTC-SVM integrated in EV propulsion system without and with stator winding fault. Simulation results have been presented in order to demonstrate that in a DTC-SVM PMSM drive, with stator inter-turn short circuits, the considered defect affects the dynamics of the motor and introduces a strong second harmonic in the motor supply currents, which can be used to detect this type of fault. This ascertainment can be used in the industry for fault detection and diagnosis of stator winding faults of PMSM.

Author details

Fatma Ben Salem
Control and Energy Management Laboratory (CEMLab), University of Sfax, Sfax Engineering School, Sfax, Tunisia

*Address all correspondence to: fatma.bensalem@isgis.usf.tn; fatma_bs@yahoo.fr

References

[1] Erdogan S, Miller-Hooks E. A green vehicle routing problem. Transportation Research Part E: Logistics and Transportation Review. 2012;**48**(1): 100-114

[2] Serra JVF. Electric Vehicles Technology, Policy and Commercial Development. London, UK: Earthscan; 2013

[3] Felipe A, Ortuno MT, Righini G, Tirado G. A heuristic approach for the green vehicle routing problem with multiple technologies and partial recharges. Transportation Research Part E: Logistics and Transportation Review. 2014;**71**(1):111-128

[4] Longo M, Foiadelli F, Yaïci W. Electric vehicles integrated with renewable energy sources for sustainable mobility. In: New Trends in Electrical Vehicle Powertrains. IntechOpen; 2018

[5] Habetler TG, Profumo F, Pastorelli M, Tolbert LM. Direct torque control of induction machines using space vector modulation. IEEE Transactions on Industry Applications. 1996;**28**(5):1045-1053

[6] Bounadja M, Belarbi A, Belmadani B. A high performance space vector modulation—Direct torque controlled induction machine drive based on stator flux orientation technique. Advances in Electrical and Computer Engineering. 2009;**9**(2):28-33

[7] Joseline Metilda A, Arunadevi R, Ramesh N, Sharmeela C. Analysis of direct torque control using space vector modulation for three phase induction motor. Recent Research in Science and Technology. 2011;**3**(7):37-40

[8] Nasri A, Gasbaoui B. A novel electric vehicle drive studies based on space vector modulation technique and direct torque. Journal of Asian Electric Vehicles. 2011;**9**(2):1529-1535

[9] Rashag HF, Koh SP, Chong KH, Tiong SK, Tan NML, Abdalla AN. High performance of space vector modulation direct torque control SVM-DTC based on amplitude voltage and stator flux angle. Research Journal of Applied Sciences, Engineering and Technology. 2013;**5**(15):3934-3940

[10] Ahammad N, Khan SA, Reddy RK. Novel DTC-SVM for an adjustable speed Sensorless induction motor drive. International Journal of Science Engineering and Advance Technology (IJSEAT). 2014;**2**(1):31-36

[11] Rashag HF, Tan NML, Koh SP, Abdalla AN, Chong KH, Tiong SK. DTC-SVM based on PI torque and PI flux controllers to achieve high performance of induction motor. Research Journal of Applied Sciences, Engineering and Technology. 2014;**7**(4):875-891

[12] Ben Salem F, Derbel N. Second-order sliding-mode control approaches to improve low-speed operation of induction machine under direct torque control. International Journal of Electric Power Components and Systems. 2016; **44**(17):1969-1980

[13] Ben Salem F, Yangui A, Masmoudi A. On the reduction of the commutation frequency in dtc: A comparative study. European Transactions on Electrical Power Engineering. 2005;**15**(6):571-584

[14] Chlebis P, Brandstetter P, Palacky P. Direct torque control of induction motor with direct calculation of voltage vector. Advances in Electrical and Computer Engineering. 2010;**4**:17-22

[15] Chaikhy H, Khafallah M, Saad A. Evaluation of two control strategies for induction machine. International Journal of Computers and Applications. 2011;**35**(5):571-584

[16] Allirani S, Jagannathan V. Direct torque control technique in induction motor drives—A review. Journal of Theoretical and Applied Information Technology. 2014;**60**(3):454-475

[17] Qi H, Li J, Chen Y. Control of electric vehicle. In: Soylu S, editor. Urban Transport and Hybrid Vehicles. IntechOpen; 2010. p. 192. ISBN 978-953-307-100-8

[18] Gieras JF, Wing M. Permanent Magnet Motor Technology, Design and Application. Electrical and Computer Engineering. 3rd edition. Boca Raton, London, New York: CRC Press Taylor and Francis Group; 26 August 2009

[19] Singh R, Sengar KP, Mishra A, Thakur C. A direct torque control of interior permanent magnet synchronous motor for an electric vehicle-design analysis total harmonic distortion of stator current. International Journal of Engineering Research & Technology (IJERT). 2016;**5**(11):57-65

[20] Hadef M, Mekideche MR, Djerdir A. Vector controlled permanent magnet synchronous motor (PMSM) drive with stator turn fault. In: Proceeding of XIX International Conference; Rome: ICEM; 6-8 September 2010

[21] Jeong Y, Sul S, Schulz SE, Patel NR. Fault detection and fault-tolerant control of interior permanent-magnet motor drive system for electric vehicle. IEEE Transactions on Industry Applications. 2005;**41**(1):46-51

[22] Ebrahimi B-M, Faiz J. Feature extraction for short-circuit fault detection in permanent-magnet synchronous motors using stator-current monitoring. IEEE Transactions on Power Electronics. 2010;**25**(10): 2673-2682

[23] Hadef M, Djerdir A, Mekideche MR, N'Diaye AO. Diagnosis of stator winding short circuit faults in a direct torque controlled interior permanent magnet synchronous motor. In: IEEE Vehicle Power and Propulsion Conference; Chicago, IL, USA; 6-9 September 2011; DOI: 10.1109/VPPC.2011.6043166

[24] Haddad RZ, Strangas EG. Fault detection and classification in permanent magnet synchronous machines using fast Fourier transform and linear discriminant analysis. 2013. pp. 99-104

[25] Kamdi PD, Vaidya UB, Kamdi SY, Asutakar P. Diagnosis of stator winding short circuit fault in permanent magnet synchronous motor. International Research Journal of Engineering and Technology (IRJET). 2015;**2**(4):325-328

[26] Rohan A, Rabah M, Kim SH. An integrated fault detection and identification system for permanent magnet synchronous motor in electric vehicles. International Journal of Fuzzy Logic and Intelligent Systems. 2018; **18**(1):20-28. DOI: 10.5391/IJFIS.2018.18.1.20

[27] Liang H, Chen Y, Liang S, Wang C. Fault detection of stator inter-turn short-circuit in PMSM on stator current and vibration signal. Applied Sciences. 2018;**8**:1677. DOI: 10.3390/app8091677

[28] Larminie J, Lowry J. Electric Vehicle Technology Explained. 2nd ed. Vol. 23. Chichester, England: John Wiley & Sons, Ltd; 2012

[29] Wang Z, Qu C, Zhang L, Zhang J, Yu W. Integrated sizing and energy management for four-wheel-independently-actuated electric vehicles considering realistic constructed driving cycles. Energies. 2018;**11**:1768. DOI: 10.3390/en11071768.

[30] Ben Salem F, Derbel N. VSC-based DTC-SVM with adaptive parameter estimation. In: IEEE 11th Int. Conf. on Systems, Signals and Devices (SSD'14),

Castelldefels-Barcelona; 11-14 February 2014; Spain

[31] Ben Salem F, Derbel N. Direct torque control of induction motors based on discrete space vector modulation using adaptive sliding mode control. International Journal of Electric Power Components and Systems. 2014; **42**(14):1598-1610

Chapter 4

Analysis of MEMS-IMU Navigation System Used in Autonomous Vehicles

Ishak Ertugrul and Osman Ulkir

Abstract

In this study, it is aimed to perform a COMSOL analysis of the inertial measurement unit (IMU), which is a sensor based on a microelectromechanical system (MEMS) in an inertial navigation system (INS). Autonomous vehicles are the types of vehicles that can go on the move without the intervention of the driver by detecting the road, traffic flow, and environment without the need of a driver, thanks to its automatic control systems, software algorithms, and intelligent sensor structures. Autonomous vehicles can detect objects around them using technologies and techniques such as radio detection and ranging (RADAR), light imaging detection and ranging (LIDAR), Global Positioning System (GPS), INS, ultrasonic sensors, camera systems, and computer vision. After providing information about the use of IMU sensors in autonomous vehicles, the finite element analysis of the gyroscope sensor by the COMSOL program was performed in this study.

Keywords: MEMS sensor, gyroscope, autonomous vehicles

1. Introduction

Recently, there has been a rapid development in autonomous vehicle technology. As a result of these developments, autonomous cars, drones, and submarines were produced. Starting with the use of electronic ignition systems and electronic speed indicators in the vehicles, the process continued with equipment such as front-rear view camera systems and parking assistants and reached the final point with vehicles moving without driver intervention [1].

Autonomous vehicles are the types of vehicles equipped with camera systems, sensors, controllers, and wireless communication modules (Global Positioning System (GPS)/INS) produced as a result of the rapid development of digital technology added to the conventional vehicles used today [2]. This equipment must be in constant communication with the satellites while the vehicle is moving. This makes it possible to determine the position of the autonomous vehicle in the world. However, they communicate with each other, control centers, and intelligent traffic signs of autonomous vehicles to determine distance and speed information, the status of other vehicles with transition priority, traffic, and road conditions of highways [3].

With satellite-based navigation systems, autonomous vehicles can quickly receive location information. These systems are used in conjunction with artificial intelligence algorithms such as machine learning, support vector machines, and artificial neural networks [4–6]. Autonomous moving, starting, and ending points are entered, and driverless vehicles which determine their route are tested. As a result of the tests performed on autonomous vehicles, very high achievements are achieved [7]. The main goal is to develop a controller that allows following the desired trajectory with the lowest error.

When we look at the history of autonomous vehicles, it can be seen that the first studies started in the 1920s. The first models that could go on their own were born in the 1980s. The first vehicle was built between 1987 and 1995 between Mercedes-Benz and the US Department of Defense with the DARPA Autonomous Land Vehicle (ALV) project. To date, all the companies producing automobiles have carried out many types of research and trials in this field.

Inertial navigation systems (INS) developed by silicon microelectromechanical system-inertial measurement unit (MEMS-IMU) sensors, which have developed rapidly in the last 20 years, are already widely used in the guidance and control of missile systems and applications developed as a basis for land and air vehicles [8]. In the inertial navigation system whose necessary calculations can be analyzed more complexly, the position is based on the measured acceleration information, and the orientation is based on the measured angular velocities.

It is seen that the systems developed with this method have been applied to various vehicles, especially in recent years, and autonomous vehicles have been developed, and highly successful results have been obtained. Examples include the determination of the direction of autonomous vehicles carrying cargo in ports, the successful use of unmanned vehicles capable of researching submarines for various purposes, the development of autonomous vehicles capable of moving unmanned in urban and intercity traffic, or the development of bomb destruction robots [9].

The inertial navigation system consists of an IMU, which is the basis of measurements, and a unit of inertial sensors, which provides information on angular velocity and linear acceleration in three dimensions. In this study, the COMSOL analysis of IMU, a MEMS-based sensor in the inertial navigation system, was performed.

2. Control equipment used in autonomous vehicles

Today, thanks to the developing sensor technology and control algorithms, various sensor-based auxiliary elements are used in autonomous vehicles with or without a driver. Some of these devices are used in distance measurement, while others are effective in determining the physical properties of objects. Besides, there is a controller or computer used to evaluate the measured data and carry out the necessary operations and other equipment that enables the movement of the vehicle. **Figure 1** shows the general block diagram of the autonomous vehicle system.

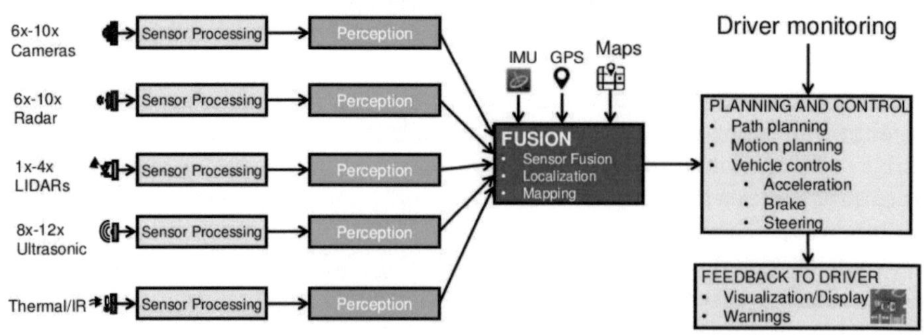

Figure 1.
Algorithm view of autonomous vehicle system [10].

The measured data from the sensors is transmitted to the control center. The sensor information is combined with the locator (GPS/INS), and the map information is transmitted to the planning and control unit. The artificial intelligence algorithm developed with this data determines the tasks that the tool must perform. These tasks are road and motion planning, vehicle speed values, braking, and management. One of the most critical features of autonomous vehicles is the display of images and warnings as feedback.

Sensors, locating, and control-based equipment used in autonomous vehicles are as follows:

- Light imaging detection and ranging (LIDAR)
- Radio detection and ranging (RADAR)
- Ultrasonic sensor
- Standard/infrared cameras and sensors
- GPS/INS
- Computer and software
- Vehicle control

2.1 LIDAR

LIDAR is a remote sensing method that uses laser or light for measuring. In summary, the LIDAR system means that it sends out a light signal and waits for the signal to return. It locates by determining how long the emitted light returns to the sensor. It transforms their positioning into point clouds of millions of points. The LIDAR sensor is expensive and requires powerful computers due to the high data flow.

2.2 RADAR

It is a system that enables us to understand, detect, or measure the speed of moving or immobile objects away from our visual range using electromagnetic waves. The system is equipped with a signal emitter and a unit that detects the signal hitting the target. When calculating the distance, the travel time of the electromagnetic beam is divided into two and multiplied by the beam speed.

2.3 Ultrasonic sensor

It is used to detect objects and measure distance with the aid of sound waves. It consists of an audio source and a receiver that senses echo. These ultrasonic sound waves have a frequency band of 20–500 kHz that cannot be heard by the human ear. With sensors using ultrasonic sound waves, distance measurements can be made without any contact with objects. The sensors are used together to transform their superiority to a more comprehensive perception. The camera system detects the color of the surface and determines the material type of the surface of the LIDAR body. When used together, the system can detect that green surfaces may correspond to grass and give the vehicle a more detailed information about the environment.

2.4 Cameras and sensors

It is used to transfer the image of objects around and inside the vehicle to the host computer in the vehicle. Unlike standard cameras, infrared cameras are used to detect the temperature of objects. Keeping cameras and sensors clean at all times is essential for data quality.

2.5 GPS/INS

They are the sensors used to locate the autonomous vehicle in the world. The GPS determines its global coordinates by receiving signals from satellites rotating around the earth. These coordinates are matched to the road map coordinates, and the position of the vehicle on the road is determined. The GPS is used in conjunction with a built-in gyroscope and accelerometer to continuously measure the vehicle's position, displacement, and speed. INS are integrated into GPS to provide location and speed data at a faster sampling frequency. These systems calculate the mechanization equations and position, velocity, and rotation angles from the IMU sensor outputs.

2.6 Computer and software

The most critical element in the use of free tools is the computer and software algorithms used in the control of the system. Interaction and communication with GPS and INS signals, traffic signs, and other vehicles are realized by a computer and software algorithm in which the controller takes place. Many of these algorithms are artificial intelligence-supported software. With this software, the information coming from the input units of the autonomous system is evaluated, and output is obtained according to the situation. As a result, the movement of the autonomous vehicle is determined.

2.7 Vehicle control

The speed of the vehicle, according to the road and traffic conditions, is achieved by the speed adjustment system. The cruise control, RADAR speed control, and adaptive cruise control are called cruise control systems. The purpose of the controllers is to automatically adjust the vehicle speed and maintain a safe distance from the vehicles in front. Steering of the vehicle is done with the help of step, servo, or hydraulic motors mounted on the steering system of the vehicle. The user decides the time and shape (remotely or closely) of the vehicle starting and stopping. Other operations (in case of emergency) are activated automatically by the computer controlling the vehicle.

3. Material and methods

3.1 MEMS-inertia measuring unit

IMU is an electronic unit that collects angular velocity and linear acceleration data sent to the central processor in a single module. The IMU consists primarily of two separate sensors. One of these is the accelerometer, and the other is a gyroscope. The accelerometer generates three separate analog signals on three axes. Due to the propulsion system and physical limits, the most critical factor in these accelerometers is that they are affected by gravity [11]. The sensor is continuously affected by gravity. As a measurement scale, there are derivatives that can measure in one, two, or three axes. These are ±1, ±2, ±4 g, and so on. It is expressed with such values (**Figure 2**).

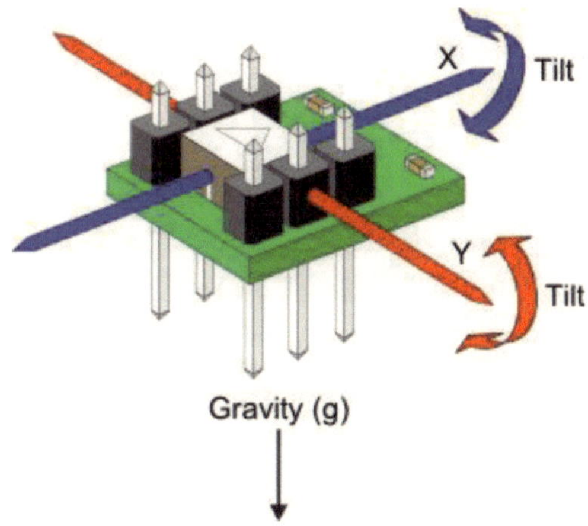

Figure 2.
The structure of accelerometer.

The second sensor inside the IMU is the gyroscope. It is formed by turning a wheel quickly around its axis. There is another circle attached at a right angle to the circle around the wheel. Another circle attached to these circles at a right angle represents the gyroscope. The gyroscope has two features. When a force is applied on the horizontal axis, a gyroscope rotating on the horizontal axis begins to rotate around the axis. Another feature is that the gyroscope remains fixed to the axis of rotation. With this feature, satellites remain earth-facing or work in autopilot applications [12]. Gyro measurements are given in radians/sec; magnetometer measurements are given in gauss (**Figure 3**).

Gyroscopes and accelerometers alone cannot provide reliable and stable data. Therefore, two sensors are combined regarding each other and information such as speed; the position is received from the IMU from a single unit. The term degrees of freedom (DOF) describes the degree of freedom of the IMU. An IMU with three-axis gyro and three-axis accelerometer is referred to as 6DOF (**Figure 4**).

When the gyroscope and accelerometer are used alone, they slip after a specific time. For example, measurements can shift 1° after 5 seconds and, therefore, cannot be used for precise measurements. We have already noted that the accelerometer is affected by gravity. When the gyroscope is used together with accelerometers that generate very high noise in the slightest vibrations, these noises are filtered. In the

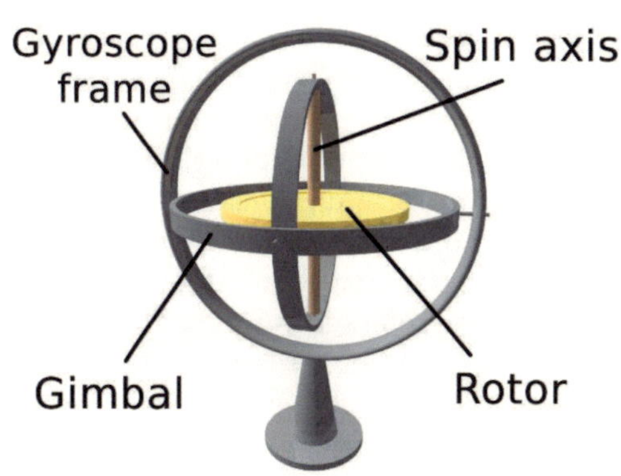

Figure 3.
The structure of gyroscope.

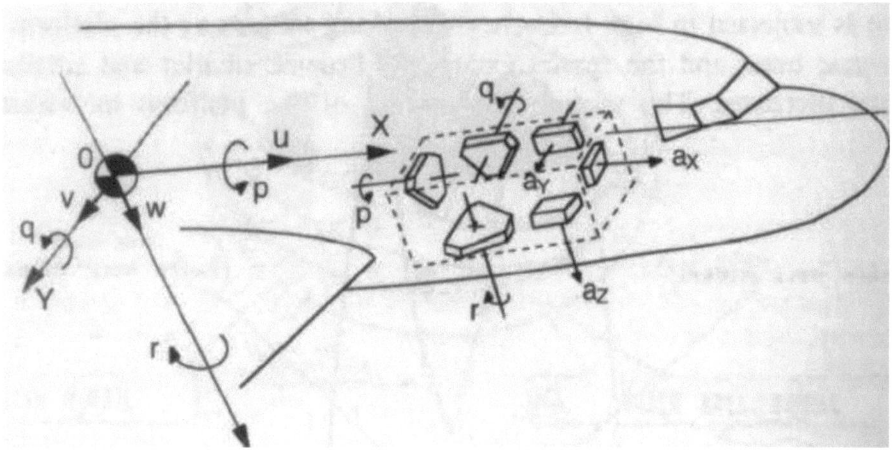

Figure 4.
The structure of IMU.

IMU, gyroscopes are used in conjunction with accelerometers as a reference since they are not affected by these forces. There are various algorithms for filtering. The most popular of these is the Kalman filter.

3.2 Inertial navigation system and GPC integration

GPS is called a Global Positioning System. The GPS sensor communicates with the GPS satellites in latitude, longitude, and altitude. Some GPS types can also provide the speed information they calculate from position changes. In autonomous vehicles, GPS data must be highly accurate and must be sampled quickly. In the measurements made with the GPS, there is a position error of up to 15 m in the 95% confidence interval, while the RMS makes a 0.1 km speed error in case of stability [13].

A GPS/INS integration is provided that provides location and speed data at a faster sampling frequency when compared to geolocation applications using GPS alone.

The IMU sensor is a MEMS-based sensor that can measure rotation speed and translation acceleration, thanks to gyroscopes and accelerometers. INS is a system that calculates mechanization equations and position, speed, and rotation angles from the IMU sensor outputs. INS cannot give accurate results as a result of errors due to noise, offset, drift, and integration in the mechanization equations resulting from the IMU sensor [8]. The INS output data should, therefore, be updated periodically with the GPS sensor using the extended Kalman filter (**Figure 5**).

To calculate the position in autonomous vehicles, the acceleration values measured with respect to the trunk axis must first be converted to the earth-centered reference system. The quaternion transformation algorithm or DCM matrices can be used for this purpose. In the quaternion transformation, the angular state of the aircraft is calculated from the angular velocities taken from the gyros. Using the obtained angle information, the transformer matrix is found from the aircraft reference system to the earth-centered reference system. This matrix is multiplied by the accelerations measured in the accelerometers. When the gravitational acceleration is subtracted from the obtained values, the acceleration of the aircraft is obtained according to the world-centered reference system.

The speed and position of the aircraft are calculated by applying numerical integration to the accelerations in the world-centered reference system. Also, since gravity acceleration varies with coordinates and altitudes, gravity must be

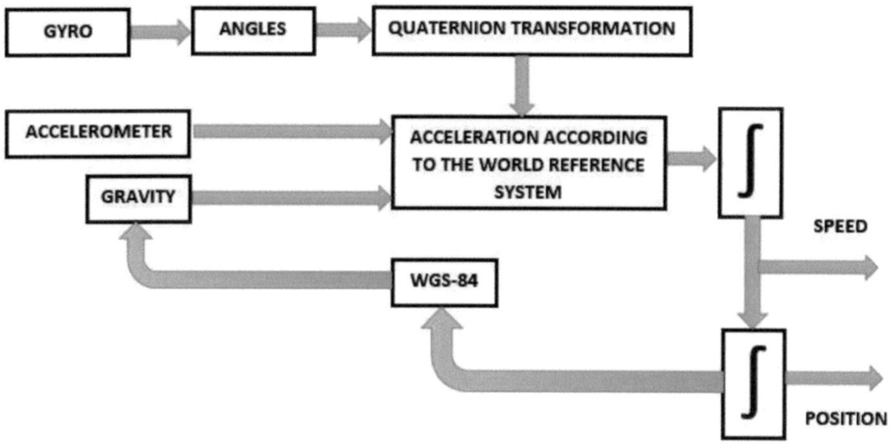

Figure 5.
The structure of inertial measurement system.

calculated based on the position information. Inertial navigation system's calculation of gravity according to a model similar to the 1984 World Geodetic System (WGS-84) is also a necessary factor for accurate measurements [14].

The inertial navigation system used in many vehicles, from missiles to submarines, requires exact and precise measurements, mainly when it is not supported by a sensor such as GPS. Also, since the algorithm used to find the position will be repeated at a high sampling frequency, the simpler the calculation, the better the performance of the system. The most important factors affecting the accuracy of measurements in gyros are drift rate and bias in accelerometers. The bias factor has a second-order effect on position calculation.

4. Result and discussion

In this section, the analysis of the gyro sensor with COMSOL is performed. According to the finite element analysis, the deformation properties of the gyro sensor were interpreted. **Figure 6** shows the design of the gyro sensor. This design was taken from the library of COMSOL, and the data were changed, and analyses were performed [15].

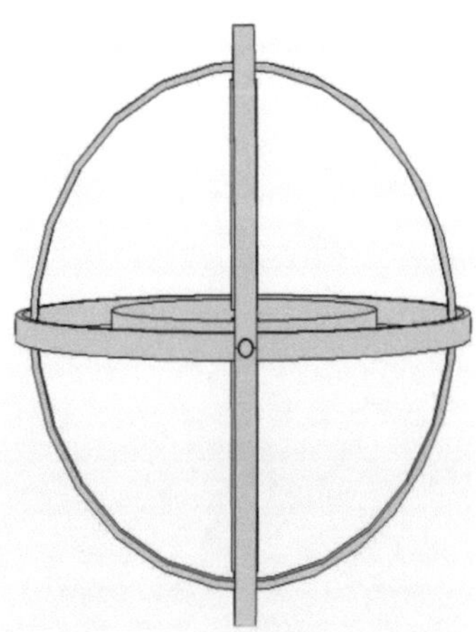

Figure 6.
Design of the gyro sensor.

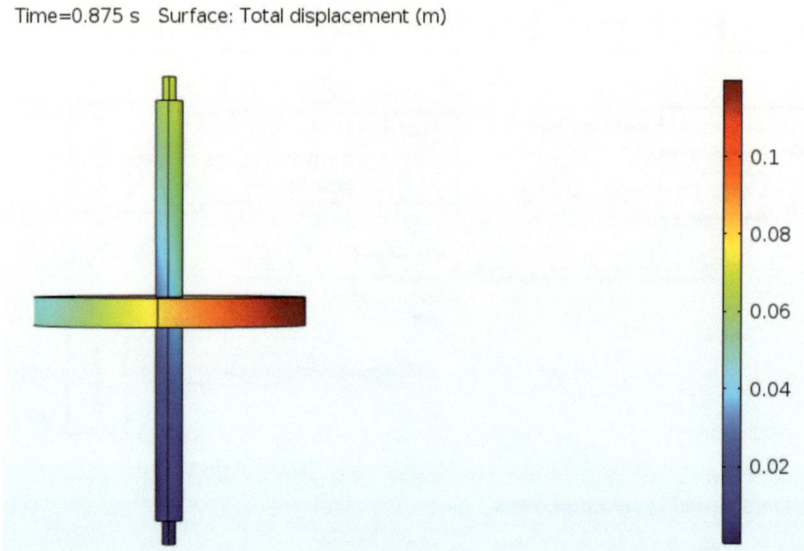

Figure 7.
Deformation analysis of the gyro sensor.

When performing the COMSOL analysis, the deformation analysis of the spinning section of the gyro sensor has been carried out because the deformation is of great importance. **Figure 7** shows the deformation analysis of the spinning section of the gyro sensor. According to this analysis, the deformation in the spinning section is a maximum of 0.12 m.

5. Conclusion

In this study, the purpose and application areas of IMU sensors used in autonomous vehicles are given in general, and the gyro sensor is analyzed with COMSOL. IMU sensors are of great importance for autonomous vehicles. Since these sensors are of micron size, their fabrication is of great importance. For the fabrication process to be successful, design and analysis must be performed correctly.

This study has brought innovation to the literature in that it provides information about the structure of the IMU sensors used in autonomous vehicles as well as about the finite element analysis.

Author details

Ishak Ertugrul* and Osman Ulkir
Department of Computer Engineering, Faculty of Engineering and Architecture, Muş Alparslan University, Mus, Turkey

*Address all correspondence to: i.ertugrul@alparslan.edu.tr

References

[1] Bas E, Tekalp AM, Salman FS. Automatic vehicle counting from the video for traffic flow analysis. In: 2007 IEEE Intelligent Vehicles Symposium; IEEE. 2007. pp. 392-397

[2] Gökozan H, Taştan M, Sarı A. Smart Cities and Management Strategies, International Strategic Research Congress (ISRC2017); Antalya; 2017

[3] Harlow C, Peng S. Automated vehicle classification system with a range of sensors. Transportation Research Part C: Emerging Technologies. 2001;**9**(4):231-247

[4] Yungang T, Jiemin Z, Yonghua H, Jianping W. A research on intelligent structure using artificial neural networks. International Journal of Pattern Recognition and Artificial Intelligence. 1997;**3**(6):1003-1011

[5] Schaab J, Muenchhof M, Vogt M, Isermann R. Identification of a hydraulic servo-axis using support vector machines. IFAC Proceedings. 2005;**38**(1):722-727

[6] Li Q, Zheng N, Cheng H. Springrobot: A prototype autonomous vehicle and its algorithms for lane detection. IEEE Transactions on Intelligent Transportation Systems. 2004;**5**(4):300-308

[7] Okutan O, Maktav D. Uzaktan Algılama Verilerinin Düşük Maliyetli Robotların Kentsel Alan Navigasyonunda Kullanımı, 5. Uzaktan Algılama-CBS Sempozyumu (UZAL-CBS 2014), İstanbul, 2014

[8] Godha S, Cannon ME. GPS/MEMS INS integrated system for navigation in urban areas. GPS Solutions. 2007;**11**(3):193-203

[9] Li Y, Dempster AG, Li B, Wang J, Rizos C. A low-cost attitude heading reference system by combination of GPS and magnetometers and MEMS inertial sensors for mobile applications. The Journal of Global Positioning Systems. 2006;**5**(1-2):88-95

[10] Available from: https://www.slideshare.net/PeterLabaziewicz/perceiving-and-understanding-the-world-for-adasand-autonomous-peter-labaziewiczv4/slide 11 of 19 [Accessed: 10 January 2018]

[11] Ahmad N, Ghazilla RAR, Khairi NM, Kasi V. Reviews on various inertial measurement unit (IMU) sensor applications. International Journal of Signal Processing Systems. 2013;**1**(2):256-262

[12] Zhang P, Gu J, Milios EE, Huynh P. Navigation with IMU/GPS/digital compass with an unscented Kalman filter. In: IEEE International Conference Mechatronics and Automation. Vol. 3; IEEE. 2005. pp. 1497-1502

[13] Godha S. Performance evaluation of low-cost MEMS-based IMU integrated with GPS for land vehicle navigation application [Master's thesis]; 2006. Available from: http://plan.geomatics.ucalgary.ca/papers/06.20239.SGodha.pdf

[14] Li Y, Dusha D, Kellar W, Dempster AG. Calibrated MEMS inertial sensors with GPS for a precise attitude heading reference system on autonomous farming tractors. In: Proceedings of ION-GNSS2009; 22-25 September 2009; Savannah, Georgia. 2009. pp. 2138-2145

[15] Available from: https://www.comsol.com [Accessed: 01 December 2019]

Chapter 5

Obstacle Detection and Track Detection in Autonomous Cars

Ayesha Iqbal

Abstract

This chapter illustrates the history and recent advancements in the field of Autonomous Vehicles with regard to two important concerns that play the most vital role in successful implementation and working of an Autonomous Car: (1) Obstacle Detection and (2) Track Detection. The car should be able to detect the obstacles for smooth and efficient working in order to avoid accident and collision. It should also be able to calculate the distance of the obstacle from the car. Similarly, Track Detection is also important as the autonomous car should stay within a predefined track and has to keep itself within the yellow lines on both sides of the road. This chapter elaborates the technologies and advancements that have been presented in the literature till date that deal with Obstacle Detection and Track Detection in Autonomous Cars/Vehicles.

Keywords: autonomous cars, obstacle detection, self-driving cars, track detection, unmanned vehicle

1. Introduction

An autonomous car is a vehicle that can guide itself without human command and control. It is also known as a driverless car, self-driving car, unmanned vehicle or a robot car. Autonomous vehicles are able to perceive their surroundings (obstacles and track) and commute to destination with the help of a combination of sensors, cameras and radars. Advanced control systems can interpret the information provided by sensors to detect obstacles and choose the most suitable navigation path for the vehicle. Enormous research has been carried out to bring the idea of autonomous car to life. Now a days, such vehicles have become a concrete reality and they have been created and extensively tested on roads, although they are not yet commercially available on a large scale.

In autonomous vehicles, one of the most important features is the correct and accurate detection of obstacles as well as the track of the vehicle. The vehicle or car should be able to detect the presence of an obstacle precisely and well in time so that it can stop itself at a safe distance in order to avoid the collision. Track detection is also a very important factor as the vehicle must be able to keep itself within the limits of track and follow the lines on the road in order to remain rightly on the track and to follow the lane as well.

This chapter deals with the study of various different approaches for obstacle detection and track detection that have been studied/implemented in the literature by different researchers. Vehicle safety systems are generally classified into types [1]:

(1) Active (DAS: Driver Assistance Systems) and (2) Passive. Active systems include collision avoidance system, automatic braking, adaptive cruise control and lane departure warning system, whereas, passive systems generally comprise features such as seat belts, air bags, crumple zones and laminated windshields. Therefore, obstacle detection and track detection falls into the category of active vehicle safety systems.

2. Sensor-based approach

The most common and widely used approach for both obstacle detection and track detection is the sensor-based approach. A number of different sensors and related technologies have been discussed in the literature. [1] discusses almost all general types of sensors used for collision avoidance that include: Acoustic, Radar, Laser/LiDAR, Optical sensors and the fusion of sensors. It also discusses their advantages and disadvantages.

Sensors can be broadly classified into two categories [2]: (1) Co-operative sensors and (2) Non co-operative sensors. They are summarized in **Table 1**.

The most commonly used sensors in autonomous vehicles are LiDAR and RADAR that are discussed below:

2.1 LiDAR sensors

In [3], a 2D laser sensor is used for obstacle detection and tracking. In [4], an automatic obstacle detection and tracking system has been introduced that fuses 3D Light Detection and Ranging (LiDAR) and 2D image data for efficient inter-distance and anti-collision management. In [5], a more sophisticated road boundary and obstacle detection scheme has been used using a downward-looking LiDAR sensor. Another study [6] presents a LiDAR and wireless sensor-based real time obstacle detection method.

2.2 RADAR sensors

Many researches have discussed use of radar sensors for obstacle and track detection [1, 7–8]. [7] uses radar sensors and uses vision and radar data fusion system for ground-based navigation. [8] uses radar and vision sensors for accurate detection of obstacles.

Cooperative Sensors	• Traffic Alert Collision Avoidance System (TCAS) • Automatic dependent surveillance and broadcasting (ADS-B)
Non Cooperative Sensors	• RADAR • Sonar • LiDAR • Electro-Optical • Infrared Sensors • Acoustic Sensors
Sensor Fusion	• To fuse the information obtained from different sensors

Table 1.
Types of sensors.

3. Camera-based approach

The second most popular approach is the camera-based approach that is used for detecting the track and obstacles in autonomous vehicles. Some researchers consider it a sub-category of sensor-based approach but due to the diversity and a broad range of camera-based detection schemes, it has been presented as a separate category.

Camera-based detection methods have been classified into three categories: (1) Knowledge based; (2) Stereo vision based, and (3) Motion based [4]. These categories have been summarized in **Table 2**.

In [9], detection and tracking of obstacles is done from a camera mounted on a vehicle with a view to driver assistance. [10] discusses tracking of people and objects with an autonomous UAV using two schemes: (1) Face Detection and (2) Color Detection. In [11], a robust vehicle detection system is described that detects vehicles in the rear view of the host car. It records the motion parameters of the host vehicle to determine the driven path. [12] introduces ROBOG, an image-based detection system, that is meant for road detection for unstructured roads.

Knowledge-based approach	• A priori knowledge of obstacles: symmetry, color, shadow, corners, vertical and horizontal edges
Stereo vision-based approach	• Using a disparity map • Using inverse perspective mapping (IPM)
Motion-based approach	• Using optical flow

Table 2.
Summary of camera-based detection methods.

4. Deep learning-based approach

There are also many deep learning approaches available for obstacle detection and tracking. In [13], one such approach is discussed which uses multiple sources of local patterns and depth information to yield robust on-road vehicle and pedestrian detection, recognition, and tracking. [14] discusses obstacle detection and classification using deep learning for tracking in high-speed autonomous driving.

5. Bio-inspired approach

A relatively uncommon and novel approach is to use a monocular camera to mimic the human behavior of obstacle detection and avoidance applied on UAVs [15]. It can be considered a sub-category of camera-based approach, but due to its novelty from the conventional camera-based approaches, it has been presented as a separate category. Similarly, [16] also discusses a bionic vision inspired approach using radar and visual information.

6. Computer-vision-based approach using PID controller

Another approach discussed in the literature [17] is a mechatronics system comprising a PID controller which predicts and controls the vehicle heading angle in order to follow the lane or to avoid the obstacles.

7. Laser scanner-based approach

Some studies present the use of laser scanner/rangefinder in order to implement obstacle detection and road following in an outdoor environment [18, 19]. This technique outperforms the commonly used camera-based vision techniques in situations such as different weather conditions (e.g. sun, rain, and fog), and different appearances of road (e.g. clay, mud, gravel, sand, and asphalt) [18].

8. Conclusion

Detecting an obstacle and keeping a track of the lane and road is one of the primary objectives in an autonomous vehicle. In this chapter, various studies and researches have been discussed that are present in the literature and focus on the obstacle detection and track detection features in autonomous cars. A wide variety of techniques discussed in the literature have been categorized and summarized according to the methodology and application of a particular technique. This chapter helps understanding and categorizing the different techniques used for obstacle detection and tracking as well as the researches that refer to these techniques.

Author details

Ayesha Iqbal
University of Management and Technology, Lahore, Pakistan

*Address all correspondence to: ayeshaiqbal_online@hotmail.com

References

[1] Mukhtar A, Xia L, Tang TB. Vehicle detection techniques for collision avoidance systems: A review. IEEE Transactions on Intelligent Transportation Systems. 2015;**16**:2318-2338. DOI: 10.1109/TITS.2015.2409109

[2] Chand BN, Mahalakshmi P, Naidu VPS. Sense and avoid technology in unmanned aerial vehicles: A review. In: International Conference on Electrical, Electronics, Communication, Computer, and Optimization Techniques (ICEECCOT). Mysuru: IEEE; 2017. pp. 512-517

[3] Habermann D, Garcia C. Obstacle detection and tracking using laser 2D. In: Latin American Robotics Symposium and Intelligent Robotics Meeting. Sao Bernardo do Campo: IEEE; 2010. pp. 120-125

[4] Aijazi AK, Checchin P, Trassoudaine L. Multi sensorial data fusion for efficient detection and tracking of road obstacles for inter-distance and anti-collision safety management. In: 3rd International Conference on Control, Automation and Robotics (ICCAR). Nagoya: IEEE; 2017. pp. 617-621

[5] Han J, Kim D, Lee M, Sunwoo M. Enhanced road boundary and obstacle detection using a downward-looking LiDAR sensor. IEEE Transactions on Vehicular Technology. 2012;**61**:971-985. DOI: 10.1109/TVT.2012.2182785

[6] Zhang J, Han J, Wang S, Liao Y, Li P. Real time obstacle detection method based on LIDAR and wireless sensor. Chinese Automation Congress (CAC). Jinan: IEEE; 2017. pp. 5951-5955

[7] Jha H, Lodhi V, Chakravarty D. Object detection and identification using vision and radar data fusion system for ground-based navigation. In: 6th International Conference on Signal Processing and Integrated Networks (SPIN). Noida, India: IEEE; 2019. pp. 590-593

[8] Kim J, Han D S, Senouci B. Radar and Vision sensor fusion for object detection in autonomous vehicle surroundings. In: Tenth International Conference on Ubiquitous and Future Networks (ICUFN). Prague: IEEE; 2018. pp. 76-78

[9] Lefaix G, Marchand T, Bouthemy P. Motion-based obstacle detection and tracking for car driving assistance. Object Recognition Supported by User Interaction for Service Robots. 2002;**4**:74-77

[10] Surinta O, Khruahong S. Tracking people and objects with an autonomous unmanned aerial vehicle using face and color detection. In: International Conference on Digital Arts, Media and Technology. IEEE; 2019. pp. 206-210

[11] Knoeppel C, Schanz A, Michaelis B. Robust vehicle detection at large distance using low resolution cameras. In: Proceedings of the IEEE Intelligent Vehicles Symposium. Dearborn, MI, USA: IEEE; 2000. pp. 267-272

[12] Kumar RK, Jada C, Feroz MGI, Kumar VA, Yenala H. ROBOG: An autonomously navigating vehicle based on road detection for unstructured road. In: International Conference on Signal Processing and Communication Engineering Systems. Guntur: IEEE; 2015. pp. 29-33

[13] Nguyen VD, Nguyen HV, Tran DT, Lee SJ, Jeon JW. Learning framework for robust obstacle detection, recognition, and tracking. IEEE Transactions on Intelligent Transportation Systems. 2017;**18**:1633-1646. DOI: 10.1109/TITS.2016.2614818

[14] Prabharkar G, Kailath B, Natarajan S, Kumar R. Obstacle

detection and classification using deep learning for tracking in high-speed autonomous driving. In: IEEE Region 10 Symposium (TENSYMP). Cochin. IEEE; 2017. pp. 1-6

[15] Al-Kaff A, Meng Q, Martin D, Escalera A, Armingol JM. Monocular vision-based obstacle detection/avoidance for unmanned aerial vehicles. In: IEEE Intelligent Vehicles Symposium (IV). Gothenburg: IEEE; 2016. pp. 92-97

[16] Wang X, Xu L, Sun H, Xin J, Zheng N. Bionic vision inspired on-road obstacle detection and tracking using radar and visual information. In: 17th International IEEE Conference on Intelligent Transportation Systems (ITSC). Qindao: IEEE; 2014. pp. 39-44

[17] Al-Zaher TSA, Bayoumy AM, Sharaf AM, El-din YHH. Lane tracking and obstacle avoidance for autonomous ground vehicles. In: 9th France-Japan & 7th Europe-Asia Congress on Mechatronics (MECATRONICS)/13th Int'l Workshop on Research and Education in Mechatronics (REM). Paris: IEEE; 2012. pp. 264-271

[18] Xu Z, Zhuang Y, Chen H. Obstacle detection and road following using laser scanner. In: 6th World Congress on Intelligent Control and Automation. Dalian: IEEE; 2006. pp. 8630-8634

[19] Wang X, Li H, Liu B. Object tracking and state estimation in outdoor scenes based on 3D laser scanner. In: IEEE International Conference on Signal and Image Processing (ICSIP). Beijing: IEEE; 2016. pp. 607-611

Section 3

Self Driving

Chapter 6

Game Theory-Based Autonomous Vehicle Control Via Image Processing

Mustafa F.S. Zortul, Tayyab Waqar and Sezgin Ersoy

Abstract

Self-driven vehicles slowly but surely are making the transition from a distant future technology to current luxury by slowly becoming a part of our everyday life. Due to their self-driving ability, they are making our travels efficient. However, they are still a work in progress as they require many software and hardware-based improvements. To address the software part of this issue, an image processing-based solution has been proposed in this study. The algorithm estimates the real-time positions and predicts the possible interaction of the objects, such as other moving vehicles, present in the vicinity of the driven autonomous vehicle in determined environmental conditions. Cameras and related peripheral present on autonomous vehicles are used to obtain data related to the real-time situation for predicting and preventing possible hazardous events, such as accidents, using these data.

Keywords: autonomous drive, game theory, vehicle control, image processing

1. Introduction

The continuous advancements in technology are constantly making our life more and more efficient. To this end, autonomous vehicles are one of the important pillars according to FIA [1] and the reports from IPSOS also show that the trends will be in the favor of the adaption of autonomous vehicles once the standardization issues are addressed [2].

Autonomous vehicles use sensors, such as radar, lidar, and camera, to implement computer vision and odometry techniques for sensing the surrounding environment and make the travel without any human intervention possible [3]. When we look into the history of autonomous vehicles, we see that first autonomous projects appear in the 1980s [4]. The first prototype came to life with navlab and ALV projects that ran by Carnegie Mellon University [4]. These prototypes were followed by project Eureka Prometheus by Mercedes-Benz and Bundeswehr partnership in 1987 [4, 5]. After these, many companies have manufactured autonomous vehicles and some of those vehicles have also found a place in active traffic in some countries. It is believed that in the near future, autonomous vehicles will cause an unprecedented change in economic, social, and environmental areas of life [6].

However, still there remains work to be done before the roads are filled with safe self-driving cars. Researches have been done to make the autonomous vehicles safer and environment friendly so that they can be as close to perfect as possible. A real-time lane detection system for autonomous vehicles is proposed by Abdulhakam A. M. Assidiq et al. utilizing the images captured by the onboard camera installed on a moving vehicle to perform real-time lane detection the lane via image processing using edge detection and Hough Transform [7]. Yen-Lin et al. presented a night-time vehicle detection method for two vehicles, incoming and outgoing, using their head and tail lights [8]. They used image segmentation and spatial clustering processes to perform rule-based vehicle detection and distance estimation by using the head and tail lights of the vehicle. Image processing techniques have also been applied to detect the current state of traffic light by Guo et al. They used Histogram of Oriented Gradient (HOG) features to implement Support Vector Machine (SVM) for the real-time detection of the current state of traffic lights for self-driving vehicles [9]. Phanindra et al. presented a method for detecting obstacles and lane using the data collected by a LIDAR sensor and a fisheye camera [10]. A vector fuzzy connectedness-based algorithm has been proposed by Lingling Fang et al. to detect the boundary of the lane using the captured images by the camera [11].

Based on these results, in this study, an image processing software has been proposed that can autonomously extract data related to factors of the vehicle environment. The predictions regarding the self-driving vehicle's environment are carried out on the basis of the proposed three models which are built using the video footage taken from an onboard camera placed inside the vehicle. The first model is an SVM that can detect vehicle positions using the inputted image. The second model fits a linear-parabolic function for the right and left lane boundaries of the vehicle. The third model approximates a bird-view version of the inputted image and carries out a more realistic approximation of steer angle on the bird-view image, rather than original.

2. Methods: research and application steps

2.1 Detection and tracking of lane boundaries

The model developed in this study is designed to work in a highway environment. Model performance is invariant to the color of the road and is directly proportionate to the amount of contrast between the road and lane boundary color, quality, and continuity of lane boundaries. However, the model assumes that the vehicle which is guided is in the center of the lane that the vehicle is present according to the horizontal plane and conducts its operations on this assumption. The rest of this subsection covers the following topics: the detection of the lane boundaries, the approximation of edge distribution function, weighted Hough transform which is used to determine the location of lane boundaries in images and lastly description of both linear and parabolic functions that we approximated for every single lane boundary. The images used in this study are taken by a vehicle-centered camera [12], and all studies are continued iteratively as a development and test loop on these images. The size of the image array (720, 1280) and different environmental aspects that it contains are shown below (**Figure 1**).

2.2 Initial detection of lane boundaries

In the first image of the application, we aimed to calculate the slopes of right and left lane boundaries in the image according to the vehicle-centered camera view. Firstly, image is converted to grayscale form and with the help of Sobel filter vertical and horizontal gradient images are obtained. These gradient images are

Figure 1.
Image samples used in the study. Images contain distortions that are caused by lane boundaries of various colors, different environmental aspects, and change of road color that is caused by various light conditions and color differences which all push the model to make errors.

then used to form a magnitude matrix and orientation matrix which contains slope calculation of every pixel in the image. With these matrices, an edge distribution function is formed, and local maxima points of edge distribution function gave us the center-lane boundary slopes on the image. Hough transform is applied with known edge slopes, and global maxima points for every edge are found with weighted voting. The locations of lane boundaries are found after slope and distance from origin calculation.

2.3 Edge distribution function (EDF)

To decrease the computational complexity, images are converted to grayscale. Function (30) is used for this conversion. The grayscale image is named as $I(x,y)$. With vertical and horizontal differentials of this function, we obtain $\nabla I(x,y)$ function [13].

$$\nabla I(y,x) = \left(\frac{\partial I}{\partial y}, \frac{\partial I}{\partial x}\right)^T = \left(D_y, D_x\right)^T \tag{1}$$

D_x and D_y represent vertical and horizontal differentials, respectively. D_x and D_y are used to calculate the magnitude matrix which contains a magnitude of every pixel in the image. Eq. (2) is used for that.

$$|\nabla I(y,x)| = |D_x| + |D_y| \tag{2}$$

After this D_x and D_y matrices are used to calculate the orientation matrix which contains orientation of every pixel in the image. In the calculation of the orientation matrix, a lookup table is used to decrease arctangent values.

$$\theta(y,x) = \tan^{-1}\left(\frac{G_y}{G_x}\right) \tag{3}$$

A voting routine in which every $I(i,j)$ pixel votes weighted as their magnitude $|\nabla I(i,j)|$ took place. After voting, a histogram of 90 bins with (−90, 90) minimum, maximum values are created.

For extracting local maxima points in the histogram, we first plotted the histogram as a piecewise function. Then smoothed the plot with a one-dimensional Gaussian filter (**Figure 2**). Filter width of 7 seemed enough to eliminate false positives after some number of tests we did. However, we also tried kernel density estimation to smooth the histogram. But it caused some meaningful local maxima points to shift from its original value or completely disappear, so we removed kernel density estimation from the workflow.

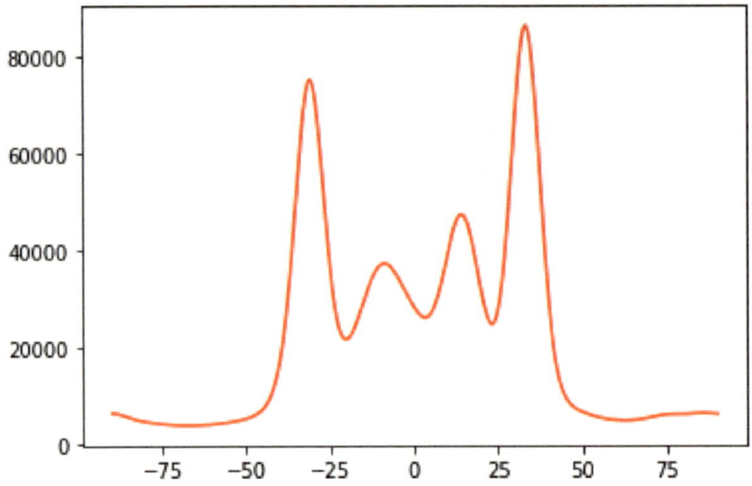

Figure 2.
Smoothed edge distribution function. Pixels between (470, 650) in the y-axis are assumed as the image part that contains the road from a (720, 1280) sized original image and converted to grayscale. Sobel filter is applied and magnitude and orientation matrices are obtained. Weighted voting took place, results smoothed with 1-d Gaussian filter.

[-30.8356546   -8.77437326 14.28969359 33.84401114]

Figure 3.
The image that edge distribution function generated from and local maxima points. Local maxima points on edge distribution function are detected and within these points, the possibility of being lane boundaries of the center lane is found as a result of thresholding.

When lane and road boundaries are considered as only linear objects in the image, we can assume that there will be two local maxima that are visibly voted higher than others. The situation on roads with multiple lanes can slightly differ from this and more local maxima points can appear in the edge distribution function. **Figure 2** above is generated from a multiple lane road. Below we discussed the local maxima points from this function (**Figure 3**).

When we consider that degrees shown above represent the degree value of horizontal axis lane boundaries, we can see that −30.8356546 is the right boundary of the center lane, −8.77437326 is the right boundary of the right lane, 14.28969359 is left road boundary, and 33.84401114 is left boundary of the center lane.

Generally, the center orientation of the center lane boundaries will be numerically closer to the vertical axis and have approximately the same absolute values.

To extract center lane boundary orientations and to eliminate false positive local maxima points which can be caused by organic variations in the road, we applied thresholding. α_1 and α_2 are minimum and maximum degree values from local maxima points, respectively. We applied thresholding to these values with Eq. (1). 15, 20, 25 are tested as threshold values and we saw that 20 shows the best performance.

$$|\alpha_1 + \alpha_2| \leq T_1 \tag{4}$$

The α_1 and α_2 couples that do not meet the requirements of this equation are removed from the local maxima list and we continue to apply the equation to the rest of the values. When there are no local maxima points left in the list, the orientation couple that closest to the vertical axis are selected as center lane boundary orientations. But it should be noted that in most cases, local maxima points of edge distribution function contain true positives only and this process is just a formality.

At this point, despite we know the orientations of center lane boundaries, we still do not know the exact locations of these boundaries. To find exact locations, we first threshold the magnitude image according to Eq. (5).

$$g(y,x) = f(x) = \begin{cases} |\nabla I(y,x)|, & \textit{if } |\theta(y,x) - \alpha| < T_2 \\ 0, & \textit{else} \end{cases} \tag{5}$$

To find the exact locations, we first get indexes of the pixels that have the same orientation as each center lane boundaries, then we zero out all the pixels except these indexes and pixels within a minor threshold value. By the end of this process, we have a magnitude matrix for each center lane boundary (**Figure 4**). As a result, we found two-lane boundary magnitude matrices and we will proceed to find exact locations of boundaries from these matrices.

2.4 Hough transform

Applying Hough transform on a two-dimensional matrix gives us a two-dimensional C (ρ, θ) accumulator matrix just as our input (**Figure 5**).

Slope and distance from image origin of found edges are obtained from the analysis of the accumulator matrix. But in this study, we already found the slopes of lane boundaries with techniques that we described in earlier subsections. Thus, we do not need to do our analysis on a two-dimensional matrix. As a result, our accumulator matrix will be independent of θ dimension and voting will take place only for potential ρ values which are only one dimension. Applying Gaussian filter

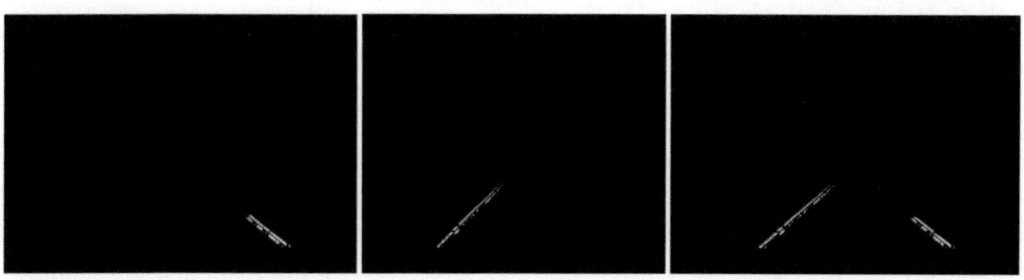

Figure 4.
Extracted edge images. These images are extracted from earlier images in this paper. A threshold is applied to a magnitude matrix to extract edge images. Two degrees is used as a threshold value in all tests. Centre lane boundaries are found with −30° (left), 33° (right). The image on the right is the sum of the edge images we found.

Figure 5.
Example Hough accumulator matrix. Hough transform is carried out as conversion of every non-zero pixel in the original image to Hough space (ρ, θ). Every pixel in the original image equals to a sinusoidal curve in Hough space. With representations of all pixels of the original image on the accumulator matrix, there will be local maxima points (ρ, θ) from straight lines of the original image. Thus, the locations of edges in the original image are found.

on these accumulator arrays will give us global maxima points. Gaussian filter windows size as 30 is enough in this part to eliminate multiple maxima points and to extract the true ρ value we are searching for which is the global maxima. After extracting global maxima points from each accumulator array, we now have the exact locations of center lane boundaries in the image. After we have exact locations, we will use these locations to specify a region of interest to be used in the next frame and edge search that we carried out will only take place in this region of interest in the next frame which will improve efficiency (**Figure 6**). Extension of 20 pixels to left and right is chosen as the width of the region of interest. We assume that lane boundary locations found in the last frame will not change drastically in the next frame of image array, this assumption is the main reason we specify a region of interest. Apart from that we also assumed that just as its location, orientation of lane boundary will not change drastically from the last frame processed to the next frame of the image array. So, we applied a threshold to the orientation search as well. We only searched in values that differ slightly from the last orientation we found. After some tests, we decided 4° as the threshold value is optimal.

2.5 Lane tracking

After the initial detection of lane boundaries from the first frame, lane detection should continue in following frames. A linear model is chosen for the initial frame.

Figure 6.
Region of interest to be used in the next frame. After Hough transform, we have exact locations and orientations of center-lane boundaries. A region of interest is specified according to this information and the process is applied only within the region of interest. The image above is the sum of the region of interest and grayscale version of the original image.

But in the following frames, we already have prior information on lane boundaries which is the region of interest. So, we can work on a more complex model to predict, a linear-parabolic piecewise function is decided as a model. For this model, horizontal line that separates linear and parabolic parts of function is chosen as approximately 15 m in front of the vehicle and area below and above this boundary named as near and far sections, respectively. The model will act linear in the near section and will act parabolic in the far section. The mathematical representation of the model is shown below Eq. (6).

$$f(x) = \begin{cases} a + bx, & \text{if } x > x_m \\ c + dx + ex^2, & \text{if } x \le x_m \end{cases} \tag{6}$$

The model needs to meet two equations to have continuity and differentiable features Eq. (7).

$$\begin{aligned} f\left(x_m^+\right) &= f\left(x_m^-\right) \\ f'\left(x_m^+\right) &= f'\left(x_m^-\right) \end{aligned} \tag{7}$$

Below equations were obtained when assumed requirements of Eq. (7) are met Eq. (8).

$$\begin{cases} a + bx_m = c + dx_m + ex_m^2 \\ b = d + 2ex_m \end{cases} \tag{8}$$

If equations in (8) are solved within themselves for c and e parameters piecewise function below is what we have Eq. (9):

$$f(x) = f(x) = \begin{cases} a + bx, & \text{if } x > x_m \\ \dfrac{2a + x_m(b-d)}{2} + dx + \dfrac{b-d}{2x_m}x^2, & \text{if } x \le x_m \end{cases} \tag{9}$$

When we assume that both right and left center lane boundaries are found in the last frame, we can expect that both right and left boundaries will be in their corresponding region of interests in the next frame. Before we try to detect center lane boundaries in the next frame threshold application below should be applied within the region of interest to eliminate various noise sources that could appear in the region of interest Eq. (10).

$$g(y,x) = \begin{cases} |\nabla I(y,x)|, & if\ |\nabla I(y,x)| > 0.5g_{avg} \\ 0, & else \end{cases} \tag{10}$$

We assume that usually, lane boundaries will have bigger values in magnitude matrices than any other source in the next frame and this assumption is the reason why we apply threshold within the region of interest. Hence, the majority of the pixels that above the threshold from Eq. (10) will belong to lane boundary we try to detect, and the majority of the pixels that below the threshold will be independent of lane boundary and can be considered as noise as a whole. As we remove these magnitudes from the magnitude matrix, we can carry out a more realistic approximation of the real lane boundary.

Let (x_{ni}, y_{ni}) be the near section non-zero pixels from threshold applied lane boundary magnitude matrices and let $\left(x_{fj}, y_{fj}\right)$ be the far section non-zero pixels from threshold applied lane boundary magnitude matrices. Also, let $M_{fj} = g\left(x_{fj}, y_{fj}\right)$ be the corresponding magnitude of matrices elements.

This situation gives us three unknown equation below.

$$\begin{cases} a + bx_{ni} = y_{ni}, & if\ i = 1, 2, \dots, m \\ \dfrac{2a + x_m(b-d)}{2} + dx + \dfrac{b-d}{2x_m}x_{fj}^2 = y_{fj}, & if\ j = 1, 2, \dots, n \end{cases} \tag{11}$$

A solution is approximated with the normal equation method for the equation above. The below function is used to represent error and tried to minimize it.

$$E = \sum_{i=1}^{m} M_{ni}\left[y_{ni} - f(x_{ni})\right]^2 + \sum_{j=1}^{n} M_{fj}\left[y_{fj} - f\left(x_{fj}\right)\right]^2 \tag{12}$$

Function E could be written as a matrix multiplication below.

$$E = (b - Ac)^T W (b - Ac) \tag{13}$$

Variables which are stated in Eq. (13) are specified below.

$$\mathbf{A} = \begin{bmatrix} 1 & x_{n_1} & 0 \\ \vdots & \vdots & \vdots \\ 1 & x_{n_m} & 0 \\ 1 & \dfrac{1}{2x_m}\left(x_{f_1}^2 + x_m^2\right) & -\dfrac{1}{2x_m}\left(x_{f_1} - x_m\right)^2 \\ \vdots & \vdots & \vdots \\ 1 & \dfrac{1}{2x_m}\left(x_{f_n}^2 + x_m^2\right) & -\dfrac{1}{2x_m\left(x_{f_n} - x_m\right)^2} \end{bmatrix},$$

$$W = diag\left(M_{n1}, \dots, M_{nm}, M_{f1}, \dots, M_{fn}\right)$$

$$c = [a, b, d]^T \ and\ b = \left[y_{n1}, \dots, y_{nm}, y_{f1}, \dots, y_{fn}\right]^T$$

To solve the Eq. (13), we isolate variable b with matrix multiplication below.

$$A^T WAc = A^T Wb \tag{14}$$

If we draw a plot with found parameters from Eq. (14), we get a straight line in near section and parabolic line in the far section as seen below (**Figure 7**).

2.6 Steer angle estimation

Steer angle estimation is prone to errors because as the distance from the camera increases the distortion of the road in the image also increases. For this reason, we first estimated the bird view version of the image with inverse perspective mapping (IPM) [14] (**Figure 8**).

We cropped the IPM image in a way that it only contains pixels from the center lane boundaries. Later we extract Canny edges from it (**Figure 9**).

After that, we apply Hough transform on the Canny image and get a two-dimensional accumulator matrix. Two highest local maxima points from the accumulator matrix give us the angle between the vertical plane and center lane boundaries. We averaged the two angle values and we found to estimate steer angle to keep the car on its lane.

2.7 Vehicle detection and tracking

We used a mix of GTI and KITTI datasets as the dataset. Dataset as a whole contains 8792 positive and 8968 negative and 17,760 examples in total. Every example is a (64, 64) size, RGB encoded, png image.

Figure 7.
Plot fitting example. Original image (left) went through the process covered in recent subsections. For found center lane boundaries, we fit a plot with the normal equation method which is covered in this subsection. Two-line image (middle) found as output and finally added to the grayscale version of the original image for comparison.

Figure 8.
Inverse perspective mapping example. Average camera parameters for bird view image of the original image (left) manually selected and bird view image (right) is found. The original image size is (720, 1280) and the bird view size is (200, 200).

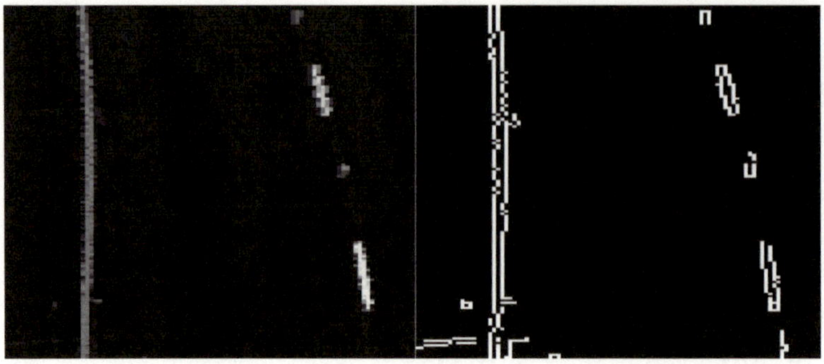

Figure 9.
Canny edges from bird view image.

First, we extracted histogram of oriented gradients in order to be used in training. Histogram of oriented gradients basically puts a grid on image. A magnitude weighted voting takes place for each bin. Then, a histogram is created from voting results.

To shrink the histogram of oriented gradients features space, the principal component analysis is applied. Then the support vector machine is trained. After training, images size of (720, 1280) scanned for vehicle detection with a sliding

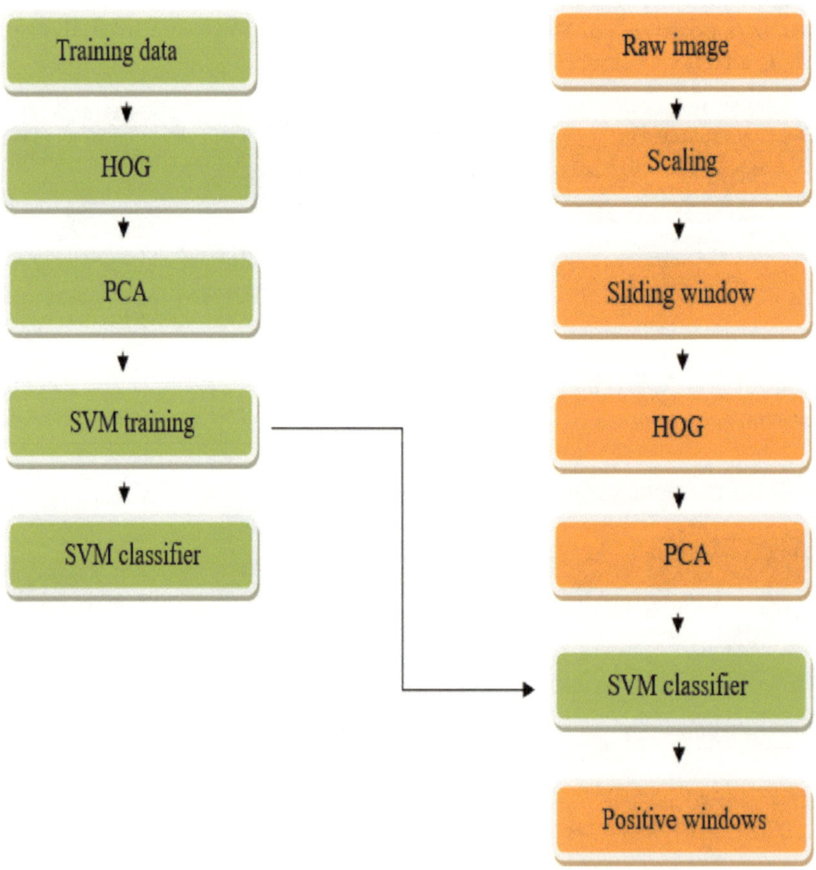

Figure 10.
Support vector machine flow chart. Green blocks represent training and orange blocks represent the test part of the application.

window. Calculated vehicle positions are used to form a heat map and the thresholded heat map is used to find a singular location for each vehicle (**Figure 10**).

3. Conclusion

This paper presents an approach based on image processing using edge distribution function and Hough transform for lane detection and tracking, steer angle estimation and vehicle detection and tracking for the autonomous vehicles. It was found that the instant change in the image feed is one of the most challenging parts in lane detection, and tracking part of this study as the model was vulnerable to instant changes in image feed. In order to prevent that, a temporal filter was applied to the region of interest which allowed shifting and hence increased the model's resistance to the aforementioned instant changes. According to the results of the tests, it was concluded that that the application of the temporal filter alone on the region of interest will not be ample; therefore, a filter was applied to the orientation of lane boundaries too, and the changes of greater than 2° in one frame. The model was found to be less dynamic, and an increase in its overall prediction accuracy was observed after those added aforementioned additions. One other issue that was discovered for this model is that it could be affected by the color changes. The model was found to be affected by the color changes on the road, shades of other vehicles, and trees on the side of the road. Errors caused by this situation usually hits the parabolic side of the estimation rather than linear.

Steer angle estimation model has a similar problem to the lane detection and tracking model. Considering the techniques that both the models share, it can be said that this was expected. This model is affected by color changes on road and shades also. If the same solution is applied, error can be lowered in color dynamic parts of the road. Another issue that was discovered was related to the parts of the road. The changes in the slope of the road cause an additional error. Because the model uses an inverse perspective mapping method that uses camera parameters as inputs, even if place occupied in Euclidean space by the camera does not change, the vanishing point which is one of the input parameters of IPM changes and this causes the model to make false predictions. To prevent this situation, image processing techniques or additional sensors can be used to estimate road slope. After predicting road slope, a model that can adapt to vanishing points shifts in the image can be developed and the error rate can be decreased drastically.

Author details

Mustafa F.S. Zortul, Tayyab Waqar* and Sezgin Ersoy
Mechatronics Engineering, Marmara University, Istanbul, Turkey

*Address all correspondence to: tayyabwaqar@marun.edu.tr

References

[1] Autonomous Vehicles. Available from: https://www.fia.com/autonomous-vehicles [Accessed: 19 June 2020]

[2] The future of mobility: Autonomous, electric and shared. Available from: https://www.ipsos.com/sites/default/files/ct/publication/documents/2019-11/the-future-of-mobility-autonomous-electric-shared.pdf [Accessed: 19 June 2020]

[3] Kato S, Takeuchi E, Ishiguro Y, Ninomiya Y, Takeda K, Hamada T. An open approach to autonomous vehicles. IEEE Micro. 2015;**35**(6):60-68

[4] Clark B, Parkhurst G, Ricci M. Understanding the Socioeconomic Adoption Scenarios for Autonomous Vehicles: A Literature Review. Project report. Bristol: University of the West of England; 2016

[5] Asadi BS, Tavana M, Asadi M, Oliver T. Autonomous vehicles: Challenges, opportunities, and future implications for transportation policies. Journal of Modern Transportation. 2016;**24**(4):284-303

[6] Azmat M, Schuhmayer C, Kummer S. Innovation in mobility: Austrian expert's perspective on the future of urban mobility with self-driving cars. In: Innovation Arabia 9: Quality and Business Management Conference (Business Innovation-Imperative for Knowledge Economy). Vol. 9. HBMSU Publishing House; 2016. pp. 142-160

[7] Assidiq Abdulhakam AM, Khalifa OO, Islam MR, Khan S. Real time lane detection for autonomous vehicles. In: 2008 International Conference on Computer and Communication Engineering. IEEE; 2008. pp. 82-88

[8] Chen Y-L, Chen Y-H, Chen C-J, Bing-Fei W. Nighttime vehicle detection for driver assistance and autonomous vehicles. In: 18th International Conference on Pattern Recognition (ICPR'06). Vol. 1. IEEE; 2006. pp. 687-690

[9] Mu G, Xinyu Z, Deyi L, Tianlei Z, Lifeng A. Traffic light detection and recognition for autonomous vehicles. Journal of China Universities of Posts and Telecommunications. 2015;**22**(1): 50–56

[10] Amaradi P, Sriramoju N, Dang L, Tewolde GS, Kwon J. Lane following and obstacle detection techniques in autonomous driving vehicles. In: 2016 IEEE International Conference on Electro Information Technology (EIT). IEEE; 2016. pp. 0674-0679

[11] Fang L, Wang X. Lane boundary detection algorithm based on vector fuzzy connectedness. Cognitive Computation. 2017;**9**(5):634-645

[12] Cortes C, Vapnik V. Support-vector networks. Machine Learning. 1995; **20**(3):273-297

[13] Lee JW. A machine vision system for lane-departure detection. Computer Vision and Image Understanding. 2002; **86**(1):52-78

[14] Bertozzi M, Broggi A. GOLD: A parallel real-time stereo vision system for generic obstacle and lane detection. IEEE Transactions on Image Processing. 1998;**7**(1):62-81